Equine Behaviour in Mind

Equine Behaviour in Mind

Applying Behavioural Science to the Way
We Keep, Work and Care for Horses

Edited by Suzanne Rogers

First published 2018

Reprinted 2020, 2021

Copyright © 5m Publishing 2018

All rights reserved. No part of this publication may be reproduced, stored in a retrieval system, or transmitted, in any form or by any means, electronic, mechanical, photocopying, recording or otherwise, without prior permission of the copyright holder.

Published by
5M Publishing Ltd
Reprinted by
5m Books Ltd
Lings, Great Easton,
Essex CM6 2HH, UK
www.5mbooks.com

A Catalogue record for this book is available from the British Library

ISBN 9781789180077

Book layout by Servis Filmsetting Ltd, Stockport, Cheshire
Printed by CPI Anthony Rowe Ltd, UK
Main cover photo by Jenni Nellist
Photos at the beginning of chapter 1, 2, 3, 4, 8 and 10 by Jenni Nellist; of Introduction, chapter 7 and 8 by Suzanne Rogers; of chapter 5 and 6 by Anthony Payne

Contents

Meet the authors		vi
Acknowledgements		x
	Introduction	1
1	Management	12
2	Breeding	32
3	Training	53
4	Equestrianism	79
5	The older horse	99
6	Riding instruction	116
7	Rehabilitation and rescue	136
8	Vets	159
9	Working animals	184
10	Human behaviour in mind	197
Index		213

Meet the authors

The Equine Behaviour and Training Association (EBTA)

EBTA consists of a group of dedicated and experienced horse owners, behaviourists, trainers and academics who have made a commitment to understanding equine behaviour especially in relation to training, and who want to promote practices with the horse at heart.

EBTA aims to:

- improve knowledge and understanding of the physical and psychological well-being of equines
- promote awareness of human behaviour and its impact on equine behaviour
- bridge the gap between academic research and practical application
- protect equine welfare whilst maintaining safety and achieving goals.

EBTA provides support to anyone wanting to learn more about behaviour, we conduct research projects in areas where we feel there is insufficient overlap between academia and the "typical horse-owner" and we liaise with media organisations to help improve communication about equine behaviour.

EBTA is currently coordinated by: Catherine Bell, Debbie Busby, Kirstin Calvin, Anna Haines, Susan Gammage, Rachel Holiday, Emily McDonald, Jenni Nellist, Jo Priede, Suzanne Rogers, Kelly Taylor-Saunders and Maisie Wake. We work in collaboration with Ben Hart, Felicity George, Amber Batson and others.

When considering suitable authors to contribute to this book, I knew that my EBTA colleagues would be a good fit because we share the same goals for raising awareness about the positive elements of the horse world and the same passion for "behaviour". This book is therefore a team effort and fits with the aims of EBTA as an organisation.

Editor (and Introduction, Chapters 1, 9 and 10): Suzanne Rogers

Suzanne is an animal welfare consultant, co-founder of Human Behaviour Change for Animals CIC, an IAABC (International Association of Animal Behaviour Consultants) certified horse behaviour consultant and registered with the ABTC (Animal Behaviour & Training Council). After a 10-year career in scientific publishing, she re-qualified in animal behaviour and welfare, gained extensive practical experience with several animal welfare organisations, worked as an equine behaviour consultant and founded Learning About Animals, running educational events. Through extensive travel to developing countries Suzanne gained an interest in transport animals and in 2005 joined the Board of the World Association for Transport Animal Welfare and Studies (TAWS). She is also co-founder and Programmes Director of Change For Animals Foundation (CFAF), and co-founder and Trustee of the Aquarium Welfare Association. In 2007, she became a Programmes Manager at WSPA (now World Animal Protection), first managing dog population and working equine programmes and later as the Technical Advisor for Human Behaviour Change Programmes. Since 2011, Suzanne has worked as an animal welfare consultant, alongside her work as an equine behaviourist.

Chapter 2: Jenni Nellist

Jenni Nellist is one of the UK's first Animal Behaviour and Training Council (ABTC) Registered Clinical Animal Behaviourists, specialising in horses, and is a full member of the Association of Pet Behaviour Counsellors (APBC). Jenni graduated from the University of Wales, Aberystwyth in 1999 with an Honours degree in Equine Science, having studied the behaviour of semi-feral hill pony mares in response to the presence of a stallion. In 2011, she was awarded an MSc in Companion Animal Behaviour Counselling from the University of Southampton, where she studied the effects of training method on the horse–human relationship. She has been helping people with horse problems since 2005, has worked in equine welfare and rehabilitation, as a stud groom, and on a livery and competition yard. Jenni contributes to veterinary journals and conferences, and volunteers for the Equine Behaviour and Training Association including providing answers for *Horse and Rider* magazine's "Ask the Expert". In her spare time, Jenni enjoys spending time with her family and relaxes by managing local Gower meadows and calcareous grassland with her Welsh Cobs, Penny and Bronwen.

Chapter 3: Catherine Bell

Catherine Bell is an equine behaviourist, certified with the IAABC (International Association of Animal Behaviour Consultants) and registered with the ABTC (Animal Behaviour & Training Council), and is an independent barefoot hoof trimmer. She therefore tends to specialise in the trimming of horses who have concerns about hoof handling. Catherine is particularly interested in maximising ethics in training and promoting choice and autonomy in our interactions with horses, necessitating resolution of the conflicting need to deliver physical care to an unwilling horse. She is also fascinated by the human behaviour element in a horse–human relationship – how do we need to "be" in order to bring out the best in the horse? Addressing behavioural problems is often more about considering how we can change our own behaviour so as not to trigger the horse, than the horse's behaviour per se. Catherine is married with two children and an elderly quarter horse, has a PhD in astrophysics and is part-way through a psychology BSc degree.

Chapter 4: Debbie Busby

Debbie is a Clinical Animal Behaviourist registered with the UK Animal Behaviour & Training Council (ABTC), and a Certified Horse Behaviourist with the International Association of Animal Behaviour Consultants (IAABC). She works with referring vets on complex behaviour problems in horses in all equestrian disciplines. After competing in a wide range of equestrian activities at riding club and county level, Debbie now concentrates on dressage and coaching. Debbie also consults in Turkey and the Middle East and takes part annually in the challenging six-day Wadi Rum desert trail ride in southern Jordan. Debbie has a first-class honours degree in Psychology and a Master's degree in Animal Welfare and Behaviour. She is a graduate member of the British Psychological Society and a full member of the Association of Pet Behaviour Counsellors (APBC), the British Veterinary Behaviour Association (BVMA), the Association for the Study of Animal Behaviour (ASAB) and the UK Centre for Animal Law.

Chapter 5: Kelly Taylor-Saunders

Kelly has been certified with the Behaviour Analyst Certification Board as a Board Certified Assistant Behaviour Analyst since 2003. She has been a qualified equine behaviourist since 2006 and is a Certified Horse Behaviour Consultant with the IAABC (International Association of Animal Behaviour Consultants). She is also registered with the ABTC (Animal Behaviour & Training Council) as an Accredited Behaviourist. She has a successful animal behaviour and training company called Solace Training (animal behaviour) and consults throughout the South East of England. She also regularly teaches workshops on equine behaviour and learning, to owners and professionals alike. Kelly has a herd of seven horses and ponies, most of which are rescued and have needed substantial rehabilitation, one of whom lived to his early thirties. With this in mind, she is well suited to write the chapter on caring for the elderly horse because this is something that she has particular interest and experience in, not only caring for her own elderly horses, but in supporting various clients over the years.

Chapter 6: Felicity George

Felicity has a strong background in science, completing a BSc and an MPhil at Edinburgh University in the 1990s, and working there as a research scientist. Her love of horses has been life-long. She has worked with horses on a voluntary or part-time basis for over 20 years; at riding schools, competition yards and rescue centres. She has three horses and a pony of her own. In 2009, she brought together her interests in science and horses and took a professional training course with the Society of Equine Behaviour Consultants (SEBC). Since qualifying, she has worked full-time as an equine behaviourist, with riding instruction from a behavioural perspective being a significant part of her work, both for riding schools and freelance. She runs a professional training course for SEBC. She is also a Certified Horse Behaviour Consultant with the IAABC (International Association of Animal Behaviour Consultants) and an Accredited Animal Behaviourist with the ABTC (Animal Behaviour & Training Council).

Chapter 7: Ben Hart

Ben hates wasted potential, so he is a trainer who uses the science of behaviour on his mission to help people understand the true nature of equines. For 20 years, he has used the science of behaviour to help both animals and their people unlock their potential. He firmly believes that working with equines doesn't have to be complicated, dangerous or stressful and by helping people to understand equine behaviour, he wants them to better understand each other and so ultimately make a better life for horses, donkeys and mules. Ben is the author of several books on equine behaviour and clicker training, as well as the creator of a unique series of individual equine training plans and online courses. He has worked with horses, mules, donkeys and people all over the world: from California to Cambodia, from working equines to racehorses. Ben's use of the science of equine behaviour rather than a one-method approach has been successful with traumatised animals at the UK's equine charities, and his ability to work with human behaviour has been utilised by leading animal welfare organisations. Ben is an IAABC (International Association of Animal Behaviour Consultants) Certified Horse Behaviour Consultant and an ABTC (Animal Behaviour & Training Council) Registered Accredited Animal Behaviourist.

Chapter 8: Amber Batson

Amber graduated from the Royal Veterinary College, London in 1999. Since then, she has worked predominantly in first opinion equine practice but developed a strong interest in animal behaviour and has been working with behaviour cases since 2001, and teaching animal behaviour since 2006. Amber was a partner in a busy veterinary practice in Surrey, UK for several years and during that time began work as a legal expert witness in animal welfare cases, which she continues to date. Amber is currently based in Wiltshire, working part-time in a local practice combined with internationally teaching behaviour and welfare principles.

Acknowledgements

Chapters 1, 9 and 10: Suzanne Rogers – Thank you to all the authors who contributed chapters of this book, everyone who provided a quote or time to be interviewed and to the whole EBTA team. A special thank you to the Cambodia Pony Welfare Organisation team for everything your project has taught me. I'd like to express my gratitude to past and present colleagues for the privilege of having the opportunity of working with and learning from you all. Thanks to my whole family for supporting the love of horses I never grew out of. Finally, I'd like to acknowledge the hard work of everyone, who in any capacity is working to improve the lives of equines with behaviour in mind.

Chapter 2: Jenni Nellist – I extend my gratitude to Suzanne Rogers and my fellow EBTA colleagues for being able to be a part of the creation of this book. In particular, I would also like to thank Dr Mark Kennedy, Mary Prewhitt, Colin Thomas, and Jackie Hughes for their contributions to the chapter, to Dr Mina Davies-Morel for inspiring in me great interest in the reproductive behaviour of the horse and passion for the humane management of this species when at stud, and to Colin Davies and family of the Cefn Stud, Gower for making it possible for me to come and study hill ponies and both admire their beauty and learn their natural ways. As ever, I also extend my thanks to my support team; partner and father of my child, Anthony Payne, and my good friend and mentor, Dr Helen Spence.

Chapter 3: Catherine Bell – I have been influenced by many professional trainers and academics, not least the co-authors of this book, as I studied behaviour. But my gratitude is particularly directed to those friends who participated in the many hours of late-night conversations about behaviour, finding our own meaning and ethical position, and who steered me off the conventional horsemanship path in the first place: Marie Eyes, Ben Hart, Rachel Holiday, Sam Jackson, Emma Kurrels, Jan Lewis, Suzanne Rogers and Julie Walsham, thank you. And of course, thank you to those particular horses who were responsible for my leaps in understanding of their needs, Muffin, Cuckoo and my elderly quarter horse Jak, without whom life would have taken a very different path.

Chapter 4: Debbie Busby – I should like to express my gratitude to Suzanne Rogers for inviting me to contribute to this book and for her support over many years in the field of equine behaviour. My thanks go to Valerie Neff who first taught me to understand horses, to Dr Helen Spence who guided me in the right academic direction and to Kelly Taylor-Saunders for her strength, understanding and humour through long years of behaviour debate and discussion.

I am grateful to Dr Emma Creighton, whose Applied Animal Behaviour and Welfare Master's degree course at Newcastle University gave me the depth of knowledge and critical perspective I needed to push my behavioural understanding to new limits. My equestrian life has been made all the richer by the generosity of Bedouin horsemen Saleem and Salem Al Zalabieh who allow me to share their beautiful Wadi Rum desert, and for that I am indebted. I am obliged to all the behaviourally-minded contributors who kindly offered their equestrian wisdom for this chapter: Sylvia Loch, Angelo Telatin, Paula Cooke, Kirsten Alexa Hansen and Megan Ellingworth. Finally, thank you my dear Badger, Flair, Rusty and Raliah; your tolerance, friendship and willingness on our many adventures gave me the deepest sense of behaviour in mind.

Chapter 5: Kelly Taylor-Saunders – I would like to express my gratitude to Suzanne Rogers for inviting me to contribute to this book, along with the many other behaviourists who I admire and appreciate. In particular, with special thanks to Debbie Busby, for her steadfast support in all things equine and life in general for the past 14 years. I am also grateful to Heather Simpson of the Natural Animal Centre where I first began my journey of applying Behaviour Analysis and Ethology to horses and at a later stage, dogs too. I am indebted to the various contributors who generously shared their experience and expertise for this chapter, bringing it to life with their personal accounts. In particular, to Dr Rebecca Giminez, who works tirelessly to ensure that large animals can be rescued safely from difficult and distressing circumstances and to Amber Batson, Debbie Busby, Damien Greenshields and Jenni Nellist. I extend love and thanks to my husband Douglas Taylor-Saunders and our daughters Scarlett and Constance, who are always behind me, every step of the way. Finally, love and thanks go to my beloved Tonka because without him, I would not have embarked on my animal behaviour journey so long ago. My remaining herd members are as much loved and I accept and appreciate all that they teach me on a daily basis.

Chapter 6: Felicity George – Thanks go first to Suzanne Rogers, for asking me to contribute to an exciting project. To Marjorie Grant, Sharon Smith and Andrew McFarlane, for their contributions to the chapter. Nicole Graham, likewise, and also for truly riding my horses with behaviour in mind, and to Eileen Gunn for 10 years of almost daily discussions on animal behaviour, and making the difficult bits fun! And huge gratitude to all the horses who have shared their lives with me and taught me – particularly to my Exmoor pony Elvis, probably the best teacher I've ever met!

Chapter 7: Ben Hart – Thank you to Suzanne Rogers, without her help this chapter literally would not have been written. May I also thank the hundreds of dedicated people working for rescue organisations across the globe, who it has been my privilege to work with and learn from. They are too numerous to mention and their tireless efforts often go unnoticed except by the lucky animals in their care; thank you all. Finally, thank you to the horses, donkeys and mules that it has been my total joy to work with for the past 20 years – they are, and always will be, the only true experts about equine behaviour and are always the best teachers.

Chapter 8: Amber Batson – I would like to thank Maruska Aylward of Bridgefield Physiotherapy and Chris Pearce of the Equine Dental Clinic for their contributions to my chapter as well as extending my gratitude to Suzanne Rogers for inviting me to be part of this book.

Introduction

Suzanne Rogers

Why another book about equine behaviour?

There are many fantastic resources available on the subject of equine behaviour. However, sometimes it can be difficult to make the connection between the theory and practice and apply what you have learnt when you are standing next to your expectant horse.

Horsemanship is changing; it has been changing since it began, and now more than ever before there is such a vast range of "methods" being promoted that it can be difficult for horse owners (and anyone involved with horses professionally or otherwise), to identify approaches that work and have compassion at heart.

Although other books on equine behaviour are available they tend to approach the subject from a theoretical angle and then include practical examples. This book has a different approach – we will investigate different elements of horse ownership and, through case studies that bring to life the ways different people have worked to meet their horse's needs, we will achieve a better understanding of what we can do to make a difference.

In recent years, there has been a move towards more "natural" ways of keeping and training horses. This development has been associated with some big-name "celebrities", who have become very well known internationally for their approaches. The field of horse training has become fragmented, with owners choosing between training methods and teachers, each of whom often has a significant "brand", and training systems that they sell and promote. This division of the equestrian market has meant that books focussed on training have an ever-decreasing audience as owners turn to one big name or system. By steering clear of promoting any single trainer and covering all aspects of equestrianism, not just training, this book hopes to be of interest to owners across disciplines and chosen training approaches.

The authors of this book support a least invasive, minimally aversive (LIMA) approach to training and behaviour modification. LIMA was first coined by Steven Lindsay (Lindsay, 2001) and has since been adopted by organisations such as the International Association of Animal Behaviour Counsellors (IAABC), through which several authors of this book are certified. Vitally, LIMA requires that trainers and behaviour consultants are adequately trained and skilled to ensure that the least intrusive and aversive procedure is in fact used. All LIMA practitioners approach a behaviour issue first by ensuring the wellness of the animal and the suitability of the environment before any training takes place.

The purpose of the book is to encourage readers to consider how typical management regimes, training and equestrian activities affect horses and how, if we understand equine behaviour, we can make changes that improve the lives of horses.

Where suggestions for alternative methods or changes are made, the aim is to motivate the reader to make positive changes for their horse or horses they work with. Care has been taken to avoid the book seeming judgemental and overly critical – however, it is likely that the content will highlight for some readers that the way horses are managed, worked and cared for often impacts their behaviour and welfare in a negative way.

Taking the good and ignoring the bad

Throughout the book, the content is illustrated by case studies, providing real-life examples and inspiration from professionals across several fields of equestrianism. We unashamedly do not support many of the things humans expect of horses and the way they are treated as objects rather than sentient individuals. However, I am personally also strongly motivated to do what I can to promote change, and for issues so entrenched in human society that poor animal welfare is considered normal, I believe that this requires an approach that embraces good things, however small, and celebrates small changes.

In searching for examples to use in this book, we purposefully ignored the bad. The stories that we have chosen are not an indication that everything about the people and the way they manage and work their horses is exemplary, we are merely including an example of one element that the person does that we think shows a behaviourally minded approach. Some would say these should not be included for fear of promoting practices that compromise welfare by association, but if we only included examples of where everything is perfect this would be a very short book. If we discounted the good examples because of the not-so-good, it would be a case of throwing the baby out with the bathwater.

Furthermore, when helping owners to support their horses towards solving "behavioural problems", we take small steps towards the change. Highlighting and building on the good things we come across is very powerful in driving further change. Often the response to suggesting changes is that resources are too limited ("I don't have a 30-acre field for my horses") but not all changes require resources, rather creative thinking. There is a case study in the human development world that illustrates this nicely (Tuhus-Dubrow, 2009). In a low-income region of Vietnam, many aid agencies had tried to address childhood malnutrition to no avail. All the typical actions were suggested, with no resulting change. One person, however, noted that even though all the families had access to very similar restricted resources, some children were healthier than others. Upon investigation, it transpired that the mothers of the healthier children were managing their resources in a way that prevented the nutritional problems the others faced – so even though they had access to the same resources as the other mothers, they were producing more balanced meals. The "successful" mothers showed the other mothers how to manage their resources in this way and mentored them for some time to ensure the new way of thinking was taken up. This was successful and finally the malnutrition rates decreased. This is a great example of managing, rather than increasing, resources and there are parallels throughout the equestrian world, whether the resource is financial, material or even compassion!

Photo: Suzanne Rogers.

A pony called Charlie

Consider a pony owned by a small city farm – a 13-hh black-and-white cob with an uncontrollable thick forelock, called Charlie, who is a firm favourite with all the children who visit. The riding school plays an important part in the community – it is situated in an impoverished area of the city with a high crime rate, high drug abuse rate and low school attendance. It is a sanctuary for many of the children who visit. Although the fence is painted with a colourful landscape showing rolling hills and bright flowers, there is no grass, no paddocks, just stables and a small yard busy with roaming chickens and ducks. The children love Charlie, he always seems happy to see them and puts his head over his stable door; he can be "cheeky" when ridden and has thrown many a child onto the ground but this is seen to be part of his personality and an endearing one at that. After their visits the children go home, draw pictures of Charlie and dream about ponies.

For the last eight years, Charlie has spent his life in a stable and is taken out for lessons six times a day, a total of three hours. Charlie spends 21 hours a day, each and every day, in a stable. He is given concentrated feed in a bucket twice a day and a haynet, morning and night. His saddle pinches him causing him to buck when it really hurts; the rest of the time he is quiet because there is nothing to do.

How should we help Charlie? Should we start a petition to close the farm for not meeting his needs? That would be one approach but likely to meet with great opposition due to the role the farm plays in the community. Should we step away because there is no grass available and no other horses to meet his social needs? Many behaviourists and trainers would do so. However, let's look for the good – Charlie is loved by his caretakers and visitors and he has played a key role in teaching compassion to many of the children who have met him. How can we work with the resources available to improve his day-to-day life? There are countless ways we can improve his life while "bigger picture" changes are worked on. We can enrich his stable to provide him with things to do in some of the time he is confined, we can change the type of food he is given and the way it is provided to better meet his behavioural needs associated with eating, we can give him time out of the stable for free exercise in the riding arena out of visiting hours, we can let him out into the yard and provide enrichment a few times a day during visiting hours to show children a different side of horses to riding them and seeing them in a stable, and there is much more we can do. We can work with the caretakers to steer their love for Charlie into activities that better meet his needs and we can enthuse them to learn more about horses and their behaviour. We can work with a saddler to address the cause of the bucking. Many of the improvements will require some training, of Charlie and his caretakers, and there might be some resistance at first from the owners and managers of the farm, but enthusiasm, knowledge of how to go about it and unbridled compassion for people and animals alike will result in creative changes that have positive effects beyond their primary focus.

As changes are seen, we are likely to have created a demand for further improvements and then we could consider bigger changes such as arranging for him to have a "holiday" of a few months a year at pasture. Small changes pave the way for big changes, sometimes rapidly, sometimes slowly, but don't dismiss the small stuff. We will explore all the ideas for changes mentioned and more in this book. We can make significant differences in Charlie's life without any change in circumstance, and although we should be mindful of the bigger goals, we can start with the small stuff right now.

What do horses need?

I have asked the question "What do horses need?" many times, in many different situations, in many different countries. The answers are never the same and always provide new insights into the complicated nature of the relationship between horses and humans. At first, people answer with the obvious response – food and water – but further probing prompts discussions about training, shade, shelter, grooming, veterinary care and much more. I often run this exercise as a framework for an introduction to welfare, so next, I ask the group to explore the criteria that should be met to fully meet that need. For example, under the need for "food", criteria might be that forage should be available as much as possible, that concentrated feed is provided in a clean bowl, that food is good quality, perhaps a specific supplement offered once a week, and so on. The group themselves come up with the needs, the facilitator just uses the questions as a framework for discussion; without having a strict idea of what the "answers" should be, the result is very much owned by the group and you can never really predict what will come up. For example, a group of pony owners in Cambodia, reportedly from the area with ponies in the poorest state of welfare, agreed that one of the things a pony needs is a name. And when discussing the criteria for this need it was agreed that the name must match the personality of the pony. In a group of livery yard owners in Surrey, one of the things a horse needs was reported to be a very precise shape of bed made from shavings. In Nicaragua, one of the needs identified was to be treated with kindness and the criteria "as much as a wife"!

As each need is identified, the group creates and places drawings depicting each one on small pieces of paper arranged in a large circle. Next, for each need, the group places a mark between the middle and the edge of the circle to represent a score for how well each need is met for the average horse in that community. If the mark is placed close to the centre of the circle, the need is not met at all and if it is placed at the edge of the circle it is well met. Once placed, the marks are connected forming one "round" of a spider's web and hence this exercise is called a cobweb analysis (Figure I.1). The resulting chart for a community where the horses enjoy a good quality of life, therefore, would have all the marks very near the edge of the circle whereas a community where the horses' needs are not met would have a round of the spider web very near to the centre. Of course, mostly the result is an irregular circle as some needs are met to a greater extent than others. The scores can be allocated considering the typical horse in a community or could be done for a specific individual horse.

The exercise is interesting for two reasons. First, it shows that people have very varied ideas about what a horse needs – but if we consider the animal irrespective of where they live, what discipline they might be used for and what culture their owners are from, what do they really need? How much of what we think they need is a reflection of the equestrian culture we have grown up in?

Second, the resulting diagram from the exercise described above, shows the whole life of a horse in an easy-to-understand way. By looking at where the scores lie, together with the criteria given for each need, we are gently guided towards how improvements can be made little by little. It is this big picture I am interested in and will keep coming back to because many of the suggestions and ideas in this book will seem tiny and almost insignificant to the reader, but when we consider the context, and imagine each of those things as chipping away to move those scores a little nearer to the ideal, we are reminded that everything is worthwhile when considering improvements to the quality of life of an animal in our care.

Please do the exercise right now for your own horse; you can add needs, criteria and assessments as you read the book and by the end you will at least have a pretty picture, if not an epiphany!

The equid ethogram: needs in a nutshell

When considering what equines need, we must introduce the equid ethogram – the term "ethogram" comes from the word "ethology", which is the study of natural behaviour. An ethogram is a list or repertoire of all the natural behaviours that horses display (McDonnell, 2003). Understanding the horse's ethogram is vital if we are to meet the natural, biological and social needs of our horses.

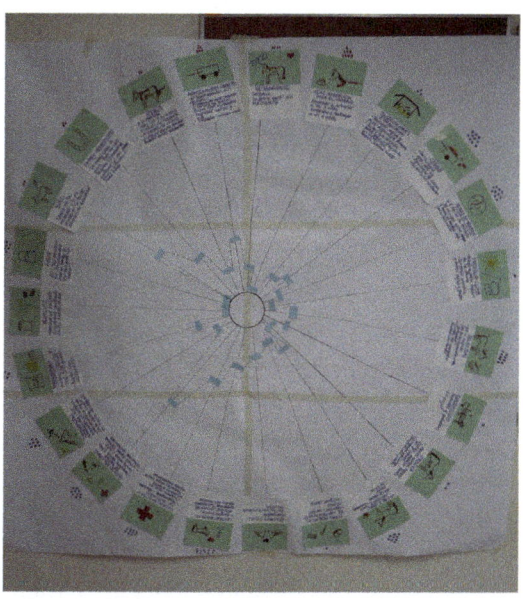

Figure I.1: Cobweb of needs. Photo: Suzanne Rogers.

Horses have a priority of needs that has evolved over 65 million years to enhance survival and reproduction. The well-known framework of the hierarchy of human needs created by Maslow (Maslow, 1943) has been modified for the horse by Heather Simpson (see Table I.1). The needs at the bottom of the triangle must be met before the other needs can be met. For example, a horse will not be able to sleep unless he feels safe. We are all familiar with this in our own lives, for example we lose our appetite when we do not feel safe, and don't feel like playing if we are sleep-deprived.

Most behavioural problems have a deep-seated cause: for example, some horses do not feel safe on their own. If this basic need for safety is not met, then behaviours such as standing still when tied up will be very difficult, if not impossible, to train. If we ensure that the horse's basic needs are met, then often the problem sorts itself out. This bottom-up approach of tackling the source rather than the symptom of the problem is known as cognitive therapy.

The needs have a time element to them. Table I.1 summarises the very core elements of normal equine behaviour and will be a useful reference throughout this book. The concepts introduced here will be built on in all the subsequent chapters. Management is covered first as this is an aspect of keeping horses that involves all horse owners one way or another; the following chapters exploring breeding, training, equestrianism and the older horse follow the life stages of a domestic horse. The chapters on teaching, rehabilitation and rescue, professionals and working animals provide a framework to discuss other ways behaviour can be taken into account.

A few words about welfare

Perhaps far into the future, humanity will have come to the understanding that even though animals might not think and feel in exactly the same way as we do, their lives are valuable, they are capable of suffering mentally and physically, and we should do everything we can do avoid that. We are not there yet. Although the term "animal welfare" is used extensively in many countries and is included in policies and legislation, it is grossly misunderstood and misapplied. For example, consider the phrase "behaviour and welfare" – this is widespread and even used as the title of some degrees and qualifications. However, welfare includes behaviour so the phrase "behaviour and welfare" clearly shows that behaviour is still not seen as being an intrinsic part of welfare.

Let's take a moment to explore what welfare really means. Imagine that you run a rescue centre, and one day you found a little alien-like creature in a box that had been dumped by your gates (Figure I.2). There is a note that says "We can no longer care for this creature; there is only one person in the world who can tell you how to look after him/her but he only responds to direct questions; you can contact him at …". So you phone the number provided and ask questions – what would you ask? Here are some examples:

- What does he/she eat? How much? How often? In a bowl? From the ground?
- What is the normal length of time he/she should sleep for?
- What does he/she like to do? Climb? Burrow? Dig? Jump? Hide?
- How can I tell if he/she is healthy?
- What does he/she do when he/she is frightened? Happy? Sad?
- Does he/she like to live in groups or alone?

Table I.1 The behavioural needs of the horse. This table shows the equine hierarchy of needs and actions that can be taken to ensure they are met. It can be helpful to consider these needs as a pyramid, with the most important at the base and higher-level needs on the next levels from the bottom up (Simpson, 2004). Higher-level needs are not addressed until fundamental lower-level needs have been met.

Behavioural need	Description	Action
Safety	Horses need to be able to feel safe and to react to threats. As social animals, this is primarily provided by keeping horses in settled, constant groups. If a horse is confined or restrained, his/her ability to respond is compromised. Horses can learn that places and situations are safe if introduced in a behaviourally-minded way. If a horse does not feel safe there is a risk of injury to the horse, humans, and other animals and objects in the environment.	• Keep horses in groups or at least in sight of other horses. • Reduce confinement and restraint where possible. • Prepare the horse for his environment and what will be expected of him using appropriate training and handling.
Eating and drinking	Horses naturally eat for 15–18 hours a day. They graze in bouts of around three hours at a time and usually take a step or two between mouthfuls. Their diet is mostly comprised of forage with 10 per cent browsing (bushes and trees). If they are working hard, or their environment lacks certain nutrients, they might need to be provided with a supplement. Providing concentrates in "meals" is contrary to the behavioural needs of trickle feeding and could mean that the horse isn't getting enough time eating, which might have behavioural and physiological implications. The amount horses drink varies widely depending on many factors including temperature and exercise. Drinking is a social activity and usually engaged in once or twice a day.	• *Ad lib* forage. • Opportunity to browse (can provide branches and twigs if not in their environment or hang hay from trees). • If concentrates are needed, feed little and often, perhaps scatter feeding. • Unlimited access to water. • Opportunity for social drinking. • Consider how water is presented; low, wide buckets are more natural than raised water provision mechanisms.
Body care	Horses take care of their coats by grooming each other or self-grooming by rubbing on trees and other surfaces in their environment, and rolling. The ability to stretch out is important for musculoskeletal health. Horses pass an average of 10 to 14 faeces a day and do not like to graze in close proximity to their droppings.	• Provide surfaces to scratch on. • Suitable substrate to encourage rolling. • Enough space to stretch out. • Paddock maintenance so that grazing does not have to be close to droppings.

Suzanne Rogers

Table I.1: (continued)

Behavioural need	Description	Action
Rest and sleep	Horses are crepuscular (most active at dawn and dusk) with polyphasic sleep patterns (sleeping, idling, dozing or resting for short periods of time throughout a 24-hour period). For REM sleep, horses must be laterally recumbent (lying on their sides). Horses prefer to sleep with a sentinel (lookout) and in groups. Sleep can be disrupted by travel, training and other activities.	• Space to allow lateral recumbency for REM sleep. • Other horses to act as sentinels. • Comfortable substrate such as bark chippings or sand. • Management and training to support appropriate sleeping habits.
Motion	Horses should have sufficient space to choose to perform essential movement behaviours that keep them fit and healthy, including walking, running and playing with other herd members. Horses play throughout their lives, geldings and stallions more than mares. There are four main play patterns across solitary play and social play.	• Keep horses in social groups. • Provide opportunity for solitary and object play. • Adequate space for exercise. • Compassionate ridden work.
Exploration and investigation	Horses should have space and opportunity through movement and object play to explore their different environments including field, yard, arena, new places and new things that are introduced to their environment.	• Provide the opportunity for investigation and exploration.
Use of space	Horses need to have enough space to retain their personal distance and to move within their flight zone. They will have a home range that they are familiar with and know their way around, and they will also explore new areas to find new grazing, drinking, breeding or other resources. Horses invite other herd members into their close proximity through communication through behaviours and body language. They thus control the direct space around them and the home range in their social group.	• Adequate space per horse in social housing and field. • Opportunity for freedom of choice about how they "use" their environment.
Association	Horses have a basic need to form bonds with other horses and feel sufficiently safe to interact with them. Interactions with other horses include sexual activity, group loafing, play, grooming and social drinking.	• Keep horses in stable herds. • Provide sufficient resources to enable social behaviours and minimise antagonistic behaviours.

Figure I.2: The alien exercise. Photo by Suzanne Rogers.

- How far would he/she naturally roam in a day?

All these questions could be clustered into the Five Freedoms, a framework for considering welfare (Brambell, 1965):

1. Freedom from hunger and thirst
2. Freedom from discomfort
3. Freedom from pain, injury and disease
4. Freedom to express natural behaviours
5. Freedom from fear and distress.

These freedoms are not hierarchical; they are all equally needed for "good" welfare. The freedoms from fear or to act out normal behaviour aren't added-on luxuries but an intrinsic part of a good welfare state. And armed with the answers to our questions we would have a much better idea of how to look after the little creature than we did before – meeting his/her physical and mental needs. It is easy to think about the things the alien might need, and usually when I do this exercise in training situations, people come up with a huge range of questions we would need to ask to be sure we could take care of the little creature well. But what if, instead of the alien, we consider the domestic horse? Are we so quick to consider all the things they need? Throughout my career as an equine behaviourist I have found that owners often consider equine welfare as mostly about health – but welfare is more than just a shiny coat. Welfare is the mental, as well as the physical, well-being of an animal and it is the mental and behaviour elements

Photo: Suzanne Rogers.

that many owners do not consider as fully as they should.

Although it is easy to get carried away by the things available in horsey catalogues, if a horse could choose what you were to buy them it would probably be different from what we would choose. Rather than a lovely new food bowl, they might prefer to have their food scattered over a clean area in front of their bed, which encourages more natural foraging behaviour, for example.

Horses only sleep for around four hours in a 24-hour period, and this time is taken throughout the day, not all in one go. However, many owners put their horses "to bed" at night, in a stable with a warm rug and enough forage for only a few hours. If we take a welfare approach we might try to maximise the time they are turned-out with friends, ensure they have access to *ad lib* forage, and design their stable with enrichment at the heart. When we think of our horse's needs from a welfare point of view we can start to think of innovative ways to enrich their life so they truly thrive; and we will consider this in detail in Chapter 1.

The Five Freedoms have been superseded by the "Five Needs" to frame welfare as elements animals need, rather than what they need to avoid. And more recently, welfare science has moved to consider quality of life: what makes a life worth living? Is an animal not just surviving but is he/she thriving? This more modern framework, the Five Domains, places more emphasis on the mental state of the animal and can be used as a welfare assessment tool (Mellor and Beausoleil, 2015). This model sets out four predominantly physical/functional domains: nutrition, environment, health and behaviour, and a fifth "mental" domain, including positive and negative emotions. A balance of outcomes across the Five Domains expresses the horse's overall "bank account balance" in terms of its physical, mental and emotional states. To enable our animals to thrive, not just survive, we must minimise negative experiences and also maximise positive experiences. The challenge is that recognition of the mental states of horses is not widespread in equestrianism, behaviour is often misinterpreted and some behaviours are often anthropomorphically labelled as positive when they are in fact negative. However, there is an increasing body of work addressing these gaps and the mental suffering of horses is increasingly difficult to ignore.

A behaviourally minded future

My vision for the future is that the currently empty words about welfare translate into actual welfare improvements, that rules and regulations that on paper should protect welfare are enforced and that "welfare" isn't just a buzzword that needs to be included to meet requirements but is valued in its own right. We will get there by building on glimpses of good welfare practice, actions that show consideration for the animals' needs, and by promoting them so that those glimpses become less transient and increasingly part of equestrian culture. Never before has the world had the infrastructure to disseminate ideas as we have now in the age of social media and web-based communication. In writing this book, the authors have searched for and found glimpses of behaviourally minded aspects of equestrianism and by putting them all together in the same place we hope that the reader will find changes they want to introduce to their lives with horses and find their place in promoting the positive changes they make and see.

We will receive criticism for promoting good things that many believe do not go far enough in changing the lives of horses. But all-or-nothing arguments just aren't helpful for driving mainstream change; the ideas in this book won't give horses a perfect life, not even close, but if ideas are acted on, seen and spread then we are edging towards real change, making positive changes in the lives of individual animals along the way.

References

Brambell, R. (1965) Report of the technical committee to enquire into the welfare of animals kept under intensive livestock husbandry systems. The Brambell Report, December 1965. HMSO, London.

Lindsay, S.R. (2001) *Handbook of Applied Dog Behavior and Training, Vol. 3: Procedures and Protocols*. Wiley-Blackwell, New Jersey, USA.

Maslow, A. (1943) A theory of human motivation. *Psychological Review*, 50(4): 370–396.

McDonnell, S. (2003) *The Equid Ethogram: A practical field guide to horse behaviour*. Eclipse Press, London.

Mellor, D.J. and Beausoleil, N.J. (2015) Extending the 'Five Domains' model for animal welfare assessment to incorporate positive welfare states. Animal Welfare, 24(3): 241–253(13).

Simpson, H. (2004) *Teach Yourself Horse: Natural horse management*. D J Murphy Pub Ltd, Surrey, UK.

Tuhus-Dubrow, R. (2009) The power of positive deviants: a promising new tactic for changing communities from the inside. *Boston Globe*. November 29.

Chapter 1

Management

Suzanne Rogers

Building on from the Introduction, this chapter will explore how well the behavioural needs of horses are met in typical management situations. Case studies will illustrate how we can address potential shortcomings even with limited resources or in constrained situations. Substantial focus will be given to the effect of feeding and environment on behaviour. Let's meet Swayze …

Swayze's day

Swayze is a 15-hh grey cob with a thin, wispy mane and a little moustache. He is owned by Rosa, a 25-year-old office worker who keeps him on part livery at a yard of 25 horses. Twenty-four hours in his life goes something like this. At 8 a.m. one of the stable girls arrives and feeds all the horses their breakfast – this comprises typical concentrate feed in a bucket. Swayze eats his food in eight minutes and then waits. At 10 a.m. he is turned out into a paddock, which used to be one large field for the herd of horses but has since been split up using electric fencing into small paddocks, about the size of four stables each. He is alone and next to a different horse each day so doesn't feel comfortable enough with his neighbours to play or mutually groom. There is very little grass, no shelter and no browsing so he stands around in the paddock for two hours until he is brought back into the stable. He isn't turned out longer as there is a rota for turnout and it is another horse's turn. At 1 p.m. he is given his "lunch", a haynet, which takes him about 20 minutes to eat. Then he stands in his stable until 5.30 p.m. when his owner visits after work. During that time, he is skipped out for five minutes, the yard cat visits his stable for three minutes and he has two short bouts of sleep of 20 minutes each. At 5.30 p.m. he is taken out of his stable, groomed, ridden in the arena for 40 minutes and then returned to the stable at around 6.45 p.m. He is given a second concentrated feed and a haynet. By 7.45 p.m., if not before, he has eaten all his food and stands around for nearly 12 hours until his day starts again at 8 a.m. He sleeps for around three hours in the night and at dawn, split up into short bursts. This is his life, day in day out.

Compared with Swayze's day, some horses receive no turnout at all, some more, some are ridden on trail rides (hacks), some just around the school, and so on but the story is common for horses across the world. What are the effects of such confinement? How can we provide more opportunity for natural behaviours, given the constraints many owners are under? And what other options are there for behaviourally minded management? This chapter will explore all these things and we meet some pretty inspirational people along the way.

The "other" 23 hours

People often ask me for my opinion on the effectiveness or kindness of different horse training methods. I do have an opinion but for me, when considering a client's horse's well-being I am concerned with their whole life. A typical amount of daily training is probably between one and two hours a day for most days – leaving a whopping 22 or 23 hours in a day, and only approximately four of those will be sleeping, and not all at once. Therefore, if we can ensure the horse's needs are met outside the training, we can minimise the impact of less than ideal training situations.

Let's consider the typical stable yard. Traditional yards in the UK consist of rows of stables, often in a "U" shape, with the horses stabled individually and turned out into fields in groups. In other parts of Europe, it is more common for horses to have access to the outside in the form of rows of stables with pens extending to the outside to provide an area the horse can freely access. These pens are usually individual extensions of the stables but sometimes more than one stable accesses the same yard area. In North America, but increasingly in the UK also, the barn system is more common, with individual stables within a large barn. Some cultures routinely tether horses as standard management practice.

Horses are confined in stables for wildly varying amounts of time but usually have access to a field at some point and this might be individual turnout or in groups. A change of scenery for a horse is usually a good thing, especially if given the opportunity to graze and browse (eat bushes and trees) but being out in an exposed field without shelter or appropriate provisions is not necessarily any better than being confined in a stable. Later in this chapter we explore many ideas for enrichment of outdoor environments as well as stables.

Horses are social animals; when living in groups, they spend all their time in visual distance of their herd-mates and most of the time they are able to see, hear, smell and touch other horses. Some time is spent playing, grooming or standing head to tail for mutual fly control. Horses need to express normal behaviours, which includes running and exercising, impossible in a stable.

Horses living in stables, especially ones that are inside and on yards that are very quiet, means that literally nothing much at all happens for hours and hours on end. The environment is devoid of any stimulation, things of interest, even of different smells and noises. From this impoverished environment, sometimes 22 hours a day or more, we expect to turn up at the yard, groom, tack up and ride our horses and for them to be absolutely compliant in doing so. Imagine being cooped up and then suddenly your world opening up as you go on a hack! That is so much to expect from our horses – when we keep them confined, their worlds become small and new experiences seem difficult to cope with due to the sudden overstimulation.

Given the increased urbanisation of many previously rural areas, idyllic equestrian properties are becoming hard to find. However, there is always something we can do to improve the lives of the animals in our care and to strive to meet their needs even within the constraints of modern horse ownership.

Meeting social needs

Being around other horses is key to survival and housing horses individually, especially when they cannot easily see or touch other horses, is chronically stressful for them (Yarnell et al., 2015). Studies in pigs have shown that when deprived of food and social interaction, pigs chose social interaction over food. To my knowledge, no such formal experiment has been done with horses, but some horses are making these choices when available in yards and are labelled fussy eaters. Eating and drinking are social behaviours for horses. Stabled horses sometimes take hay from their haynet, which might be in the back of their stable, and walk to the front to eat it looking over the door where they can see other horses. This could be for several reasons, including a need to see what is happening outside if they do not feel safe eating alone, but in many cases is likely to be rooted in the desire for social eating.

Group housing can help to meet the need for social interaction if confinement is considered vital. Horses can be kept in small groups loose inside a barn or in a sheltered area with hardstanding. The size is important and must provide enough space for horses to move away from each other without getting cornered. Interestingly, keeping horses this way has been found to reduce the time it takes to muck out compared with horses kept in individual stables.

Social housing works best if the food and water points are not placed in the corners of the area – because eating and drinking are both social activities. Having to take turns because a drinking trough is only accessible by one horse at a time can be stressful and cause negative interactions between members of a herd. It is important to introduce horses in a large area first before decreasing the area available. Many horses have not developed social skills that enable them to live in small spaces with other horses, so a gradual introduction plan is vital, ensuring that resources are plentiful.

One of my clients kept her horse on a yard where the stables were in a barn which was divided into many "stables" using relatively low, concrete walls. At first, the design could be considered appealing due to the fact the horses could see, touch and smell other horses on three out of four of the sides of their stable. However, upon observation, it was clear that this was actually making the horses very anxious – perhaps partly due to being able to see a large area but not have access to it but also because of the interactions between the horses. One of the horses was frequently lying down, which the owners read as a sign of contentment with the environment given that, as prey animals, lying down is a vulnerable position. However, in this case it seemed that the horse was lying down as it was the only way he could remove himself from contact with other horses on three sides of the stable. This is a good example of where an owner was trying to think from the horse's perspective but lacked the observational skills to be able to assess the horse's likely emotional state.

Designing stables with "windows" between them enables horses to touch and smell each other but the windows should be relatively small so that the horses have their own space to retreat to in case of any negative interactions. If the stables are small and the horses have not learnt appropriate social behaviours, it might be very stressful for a horse to have such contact.

An innovative approach to social housing is the use of wall systems – walls that stand independently and are moveable, placed in the middle of social housing (to avoid creating corners where horses could get trapped) to provide visual breaks from other horses when needed.

For horses stabled with no tactile access to other horses, a "visit" from a friend can be a welcome activity. Consider leading another horse or pony outside your horse's stable so that they can interact. If they know each other well then supervised time with the visitor horse tied up outside the stable might provide positive experiences for both. The choice of visitor is important, ideally it should be horses who are

Photo: Jenni Nellist

turned out together and who are known friends. If wanting to introduce horses that can visit each other safely, they should be introduced using a standard introduction programme before being placed in close proximity to each other.

Exploring the use of stable mirrors

There is a growing trend to use stable mirrors as a form of enrichment, ostensibly for the horse to see "another horse". This concept needs to be explored. First, it is unlikely that the horse really interprets the mirror image of himself as being another horse – the "other" horse will lack the smell, sounds and behaviour a real horse would have. Even if the horse did interpret the horse to be another horse, that might not be a positive thing if the horse was exhibiting signs of stress or anxiety – perceiving a stressed horse in the same environment is unlikely to be a positive outcome for the real horse. There is the possibility that the horse will recognise the horse in the mirror as him/herself. Until the last decade or so it was thought that only humans have the capacity to recognise themselves in mirrors. In 1970, a scientist called Gordon Gallup Jr devised a methodology to test whether a non-human animal has the ability for self-recognition. Scientists first introduce an animal to a mirror and observe their behaviour. Then they place a visual mark on the animal's body (e.g. scentless paint), introduce the mirror again and observe any differences in behaviour compared with the times the animal was in front of the mirror without the mark. Animals that pass the test and are thought to understand that the animal in the mirror is themselves, adjust their bodies to get a better look at the mark using the mirror and might touch it (on their body, not in the mirror). So far, only a few species have passed the test, not including horses. However, the test is criticised for producing false negative results for many reasons – for example, typically low numbers of animals are tested and the individual tested might not be representative of their species, and if kept in an impoverished environment they might lack the curiosity they would otherwise exhibit. With the current scientific knowledge, it seems that horses do not recognise their reflection as themselves, but this requires further investigation. Ultimately, however, we still don't know if horses consider the horse in the mirror to be another horse, themselves, a representation of a horse, something else entirely or if the perception varies between individuals.

The research appears to be inconclusive regarding the results of the use of stable mirrors; some studies suggest that they help to decrease stereotypical behaviour (McAfee *et al.*, 2002; Mills and Davenport, 2002). To introduce a mirror, it is best to first test your horse's reaction to the mirror in a larger space than the stable – or in the stable but so that he has the option to freely enter and leave. The general advice is that the mirror should be placed in the stable so that the horse can choose whether to look at his/her reflection or not. However, this possibly fails to take into account the nearly 360-degree vision of equines.

The initial response of horses faced with a mirror varies significantly. Some horses approach it and extend their noses in a typical "social greeting" behaviour, some even make vocalisations directed at the mirror and some horses make aggressive threats to their reflection (possibly more likely in undersocialised animals). Studies, and general observations have shown that many horses are cautious when they first see the mirror but that they "accept" the mirror and benefit from the reflection after a couple of weeks. This is interesting as has the horse learnt in that time that the reflection is not a horse, or are they considering the reflection to be a horse?

Movement

Free-ranging horses continually travel to find food and water. It is estimated that horses roam for around 32 km a day given the opportunity and, interestingly, the number of steps taken is similar in large and small areas. The physiology of the horse is set up for this – horses have relatively small hearts for their body size so movement aids blood flow around the body. When stabled, however, the circulatory system is not able to work this way and the heart does not work optimally.

Movement is also needed for optimal functioning of the skeletal system, especially in development, as it stimulates growth in bone density. And movement over natural terrain stimulates the hooves to grow and be worn down appropriately. Dr Robert Bowker, an equine hoof researcher, showed that horses who were stalled 24/7 only took 800 steps per day compared with horses living in fields, who took 8,000–10,000 steps per day (unpublished). Interestingly, he found that even stabling horses just at night resulted in less movement during the day even if given the opportunity.

Although horses tend to have more freedom in a field or paddock than in a stable, the growing trend to split fields into many sections to provide horses with individual turnout, sometimes not much bigger than a stable, is of great concern. The size of the home range for free-ranging horses varies according to the resources available on the land but is estimated to be up to 78 km^2 (Mills and McDonnell, 2005). When we keep horses in small fields and deny them the opportunity to home range it is our responsibility to meet their behavioural needs in other ways.

Another welfare concern is that horses do not like to graze around faeces, either their own or that of other members of the herd. When kept in small fields, horses usually develop the habit of eliminating in one or two places (termed latrines) and as these are not grazed it means the grass is taller in the toilet patch. Free-ranging horses do not develop latrines unless their home range is small (Lamoot et al., 2004). For good management in establishments where grazing areas are small, the faeces should be picked up so that the horses have a larger area to graze. In one experiment, when grass was cut from a clean area, and from an area used for elimination, horses refused to eat both sets of grass if a faecal bolus was present, indicating that it is the presence of faeces that prevents the horses from eating that grass and not any difference in palatability of the grass itself (Ödberg and Francis-Smith, 1977).

There is some evidence to suggest that inappropriate stabling (where the horse does not have the choice between confinement and more freedom) alters hyperopic (far-sighted) vision and induces myopia (short-sightedness). We could be creating short-sighted horses by stabling them, especially during development (Gumm, 1991) – perhaps in the future, horses will wear glasses to correct their vision!

If turnout is limited, horses can be given the opportunity for free movement by being turned loose in an exercise arena. If done with another, known, horse this also provides an opportunity for social interaction. You could place objects for the horse(s) to investigate such as toy buckets or cones around the arena, taking care not to include items the horses might find frightening.

While we explore how to meet the need for movement, let's consider the ethics of riding. My opinion with respect to my own horses is that I keep them on about 10 acres in a herd of four. They have a relatively enriched set of fields, with trees to browse, hedges, various options for shelter and good grazing. But this is still far from the home range they "should" have. My horses are bitless, treeless, shoeless and were trained using positive reinforcement, building up to hacking for hours. When hacking I sometimes ride, sometimes walk beside them. I let them browse trees and stop for a "munch" on grass periodically. The indications are that they enjoy it – the area they get tacked up in is pretty much only used as a precursor to being

Photo: Jenni Nellist.

ridden and yet when I haven't ridden for some time they are very keen to enter that area. Could the rides they do regularly be considered part of their home range? On balance, I believe that currently riding my horses is a positive part of their life but one that I reconsider regularly and I find myself increasingly going out with them in-hand rather than under saddle. I realise that the sentiment above is very different to the everyday equestrian, and also doesn't go far enough for some people but for me, with behaviour in mind, for my horses in our situation I am comfortable with that decision.

Considering that horses move mostly while they are eating, "patch foraging" as they go, now that we have considered how to meet the need for movement, we should next consider how to meet the behavioural needs associated with eating.

Towards natural eating behaviour

As well as the nutritional make-up of the food we provide our horses, we need to consider the way we provide it. Horses are typically fed from buckets and haynets but with some creativity we can present feed in a more naturalistic way.

Browsing

Horses naturally graze on ground-level grasses, small plants and herbs, and browse on shrubs and trees. Horses like to take at least 10 per cent of their diet as browsing but this is rarely the case for our domestic horses in modern times. Eating twigs, tree bark and even wood is normal equine behaviour. Horses especially like oak (no acorns), ash and poplar trees; and apple branches are also popular. If you have access to woodland or trees, then small-scale harvesting of twigs, leaves and branches that can be tied up in the stable can allow browsing behaviour in the stable or field if not available to the horse otherwise. Perhaps create a "treat branch" by adding fruit or vegetables, skewering raisins or other dried fruit to twigs sticking out of the branch, or smearing pureed fruit or herbal pastes to encourage exploratory and foraging behaviour. Or create a rope plait, tie it up in the stable or to a tree, and poke through carrots, hay and other treats (see Figure 1.1).

Ways to replicate browsing on trees is to take some forage such as low-calorie haylage or hay soaked in flavoured water or even plants such as dried nettles and place them in small-holed nets hung at heights that encourage the horse to stretch up slightly to reach them. Sometimes I create "hay trees" by putting hay in trees for the horses to reach up and browse. Alternatively, small edible branches, such as hazel, beech, ash or willow, can be tied up in the corner of the confined area for leaves to be browsed and bark to be stripped. Once stripped, they provide a framework for smearing on tasty pastes such as low-sugar jam, herb pastes, yeast extracts or even threading on low-sugar mints, small pieces of bread, pieces of low-sugar vegetable including swede, cabbage, cauliflower, and so on.

In 2005, researchers offered horses and ponies 15 different flavours, of which eight were favoured (Goodwin *et al.*, 2005). These are cherry, cumin, fenugreek, peppermint, carrot, oregano, rosemary and banana. Interestingly, four of the other seven flavours all commonly fed to horses – apple, garlic, ginger and turmeric – were accepted but eaten much more slowly than the top eight flavours. The remaining three flavours tested, echinacea, nutmeg and coriander were not universally accepted.

Providing forage

There are many different types of hay and haylage available and even bales from different

Figure 1.1a-b-c: Food-based enrichment. a, place pony treats in the boxes, which they tip up to get the treat; b, browsing rope with cow parsley and carrots; c, different herbal teas to smell or soak hay in. Photos: kindly provided by Alice Robinson.

fields of the same plant have slightly different qualities and will provide interest. Therefore, a simple way to add some variation into your horse's diet is to feed different types of hay and haylage (Goodwin *et al.*, 2002). You could create flavoured hay by soaking some in a bucket of herbal tea (e.g. chamomile, peppermint); one teabag is suitable for a bucket of water with hay in if you poke/prod it quite a lot to infuse the water before adding the hay.

Placing the bulk of the forage on the stable floor can be done in a way that takes into account that horses like and need, to eat and move. If stabled, placing forage along the walls, rather than just in one corner, encourages gentle moving and eating; this is appropriate for almost all horses except those with a condition requiring total immobilisation. For field-kept horses fed supplementary hay, it should be placed in many piles, not just the required "one per horse plus one" rule to avoid conflict over perceived restricted resources.

A kilogram of hay is consumed more quickly than the same weight of fresh grass. Therefore, feeding such forage in ways that slows down the rate of consumption will help to make it last longer. The usual way to feed hay and haylage is in nets. The use of haynets has been shown to increase feeding time compared with when fed from the floor, so could be considered for a portion of the daily forage ration each day (Glunk et al., 2014). There is a growing movement to use nets with small holes so that the horse finds it more difficult to access what is inside, takes longer to eat and therefore provides more time-consuming forage. However, some research shows that it can be incredibly frustrating for horses to not be able to access the feed easily and some animals give up before they are sated. To address this, pack the hay loosely in the net and provide a mixture of easily and less accessible hay. In time, you will be able to shift the balance so that an increasing proportion of the hay is fed in a net, as the horse gets used to eating that way. Feeding from the ground rather than a haynet has additional advantages; some horses are allergic to dust and spores that can be present in forage and this is made worse when the horse eats food that is raised.

Straw beds can also be useful as horses will top up their "grazing" requirements from this low-calorie bedding, however it must be considered that the lack of movement in the 24/7 confined horse does predispose them to the increased risk of impactive colic. One study found that horses bedded on straw spent the most amount of time lying down and showed the least "behavioural problems" compared with some other bedding materials (Kwiatkowska-Stenzel et al., 2016). One study found that the amount of agonistic behaviours reduced when group-kept horses had access to straw, suggesting that freely enabling forage behaviours was key to preventing behaviour that does not routinely happen in the wild (Jørgenson et al., 2011).

In Switzerland, new automated feeding systems have been introduced at livery yards, private properties and the National Stud. Some are electronically operated – owners fill the hay container and then set the machine to open and close periodically over a 24-hour period, so the horse never goes more than two or three hours without access to forage. Further studies are needed to assess the psychological effects of these systems – for example, do some horses spend their time watching the container waiting for it to open or are they able to use the time without access doing other behaviours such as resting?

Hard feeds

Typically, most horses are fed "hard feeds" as well as forage – processed food that although often providing the required nutrition, does not provide the appropriate chewing action or behavioural element of grazing and browsing. Different horses have differing nutritional needs and it is worth engaging the help of an equine nutritionist to work out a diet suitable for your individual. However, in general terms, chopped forages such as straw, alfalfa and grass, take longer to eat than more processed feed. You could consider using small piles or buckets with different parts of the feed in each one, so the horse can choose which to consume first. This can be fascinating to watch and will take the horse a little longer to finish than if all the feeds are mixed together in the same container. Scattering feed over the ground (as long as it is suitably clean) at the front of the stable instead of feeding from a bowl can also help to slow down intake and provide an opportunity for more active forage behaviour.

Other food-based enrichment

Root vegetables such as carrots, swede and parsnip can be chopped up and included in the feed, mixed into layers of forage or even hung up on baling twine. You can plait ropes, poke sliced vegetables through the holes and hang the rope up to simulate browsing behaviour.

Growing herbs in pots or tyres around the yard can be aesthetically appealing and horses sometimes enjoy smelling and eating them. To avoid too much damage, a metal grid can be bent over the plants so that the horse can't eat your carefully nurtured plants in one mouthful.

Some horses particularly love apple bobbing. The same as with humans, you float an apple in a bucket and the horse tries to bite the apple but it bobs out of the way. You need to teach this game in steps, first with just a centimetre of water in the bucket, adding more in time until the horse has learnt the game. Some horses and ponies solve the problem by dunking their heads to get the apple! This behaviour could simulate solitary object play behaviour, food acquisition, problem-solving or investigative behaviour.

More "off-the-shelf" enrichment ideas include "licks", which are flavoured blocks that the horse can lick. Sometimes they are mineral-based but often they are made from sugar so are not suitable for all horses. You can purchase lick holders that swing, rotate or move in some way to provide a little more stimulation. Ball feeders are readily available – you fill them with concentrated feed and the horse pushes them around making the feed fall out. These need to be introduced gradually: first, place some food on the floor next to the ball to encourage the horse to investigate and you might need to scatter food over a small area with the ball to help your horse to learn to push the ball for food. You can make your own version of a ball feeder using a sturdy plastic bottle (without the lid). Cut holes large enough for the food to fall out of the bottle, fill with food and allow the horse to use it under supervision.

We must bear in mind that edible, tasty treats that aren't instantly accessible can be frustrating for some horses and instead of providing enrichment provide a source of anxiety, frustration and stress. Also, if your horse tends to guard food it is important to address that issue before introducing devices such as those described.

In summary, feeding our horses with behaviour in mind involves a need to think innovatively about ways that we can meet not only our horse's nutritional requirements but also their behavioural requirements regarding eating.

Non-edible enrichment

There is a variety of non-edible enrichment items available on the market (see Figure 1.2).

Examples include:

- Rubbing mats: textured rubber mats can be purchased that can be attached to the stable wall for your horse to rub against. Some horses love to use these in the summer months especially. Naturally horses would groom each other, or groom themselves on surfaces such as trees, so this is meeting a potentially unmet need for grooming. Home-made grooming spots include attaching broom handles to the stable wall or a post with multiple broom handles attached to a yard area. The "Pillow Post" (created by EBTA's Emily MacDonald) is an ingenious pillow case made from the same material as rugs for horses, which you tie to a post to provide a soft surface to rub on and can be used in stables or outside.
- Toy box: some horses enjoy investigating different smells and textures and a good way of meeting this need is to introduce a toy box – a large, flexible bucket filled with things such as cut-open fruit juice

Figure 1.2 a-b-c: Non-food-based enrichment. a, brushes for scratching on; b, novel object to investigate (a and b kindly provided by Alice Robinson); c, a "Pillow Post": this can be used as a safe rubbing surface for horses. Photos: a and b kindly provided by Alice Robinson, c kindly provided by Emily McDonald of Meadow Family Rescue who makes the product.

cartons, soft toys on the market for dogs, boxes that have contained coffee grains, towels, brushes and anything else that is interesting and safe. The toy box will provide an opportunity for solitary play and exploration.

- Stable toys: one study investigated the long-term interest of horses in ropes and plastic hanging balls (Bulens et al., 2015). Researchers found that horses did use the items but to a limited extent and there could be many reasons for this including: once the animals had investigated the objects, there was no need to do so again to fulfil the need for exploration; the horses did not feel safe enough in the environment to perform higher-level behaviours such as exploration; or the horses had "enough to do" in their stables as those studied had good access to forage.
- Novel objects: presenting novel objects for horses to explore is enrichment. For example, a big cuddly toy (safe for babies with no dangerous parts), an inflatable, a

tarpaulin or other floor covering or even a paddling pool!
- Rolling: another behaviour we can encourage through the design of our horse's habitat is rolling. Horses prefer to roll in soft, sandy substrate so providing a sandpit area would be great to do if feasible.
- Routine tying: where horses are stabled for long periods of time, a change of scenery can provide some variety. If we tie a horse outside their stable, using rope that is long enough for the horse to move their head to each side but short enough to avoid getting tangled between their legs, we provide a different perspective for them and also provide a learning opportunity – they might be able to see vehicles, wheelbarrows, flapping bags and people and as long as the horse is not fearful of these things it will provide valuable exposure to a variety of stimuli and therefore develop a foundation of experiences they have the ability to cope with. Being given the opportunity to eat from a haynet might also meet the need for social eating if tied in view of another horse – not so close that the horses become anxious.
- In-hand walks: taking horses out for short walks in-hand can provide a change of scenery and possibly the chance to graze for a few minutes on any grass verges. Standing next to your horse while they are grazing provides the opportunity for owners to just "be" with their horse, without the horse under the control of being ridden or an active training session. The horse will associate your presence with a positive activity. If there are no verges, you could prepare an area where chopped vegetables and pony nuts are scattered and take horses to "graze" the area with the scattered food.
- Consistent carer: if a yard has paid staff to care for the horses, it is worth considering having the same handler for a specific horse most days. This provides the opportunity for the horses to have relationships with people, and as the horse spends more time with the same handler they each learn the other's habits and preferences. Each staff member should be responsible for the enrichment of the horses in their care as part of the key care activities. Alternatively, a member of staff could be employed solely for enrichment, a role as vital as other caregiving roles ensuring that it is not pushed aside by daily pressures.
- Tea breaks: horse owners on sociable yards often drink tea or coffee during the time they visit their horses. Time having breaks from chores or eating lunch can be well spent just being near your horse or watching horses interact in the field. You might be observing your horse, but your horse will also be observing you!
- Field enrichment: if you keep your horse living out in a field, enrichment is still needed as it is likely that the field is smaller than a natural home range. Very few of us have the luxury of keeping horses in tens of acres of land, with varied terrain, plants and trees, and so on. Fields can be enriched by offering varied water sources, sandy rolling areas, rubbing posts, logs or poles to walk over or navigate around and artificial hills; many of the ideas for stable enrichment are valid too.

Embedding enrichment

Enrichment should not be considered as a luxury or an optional extra, it should be part of standard equine management and taught by all people-training establishments. It is a vital part of helping horses to cope with the lives we can offer them and, just as many horse owners aim to keep their horses physically healthy, we must give the same attention to mental health otherwise their welfare must be considered compromised.

Consider keeping an "enrichment chart" with what you have done or plan to do that day or week. If you are creative you could design a chart – one of my clients cut out horseshoe shapes, split them into seven sections, wrote a daily enrichment plan in each section and displayed it on the wall outside her horse's stable. Another yard I visited was inspired to truly incorporate enrichment into their routine and wrote it on the feed-board for the horses, giving enrichment the same priority as nutrition.

Nutrition and behaviour

Now that we have considered how to meet our horses' needs concerning the social elements, movement and eating, this section covers the fascinating links between nutrition and behaviour. Dr Andrew Hemmings, a lecturer at the Royal Agricultural University in the UK, said, "The primary cause of abnormal behaviours in the domestic horse is that we fill them up with the equivalent of Lucozade or Red Bull and then keep them in a stable." To be equipped to fully understand this, let's consider the evolution of the horse. Around 65 million years ago, the first equine ancestor appeared – Hyracotherium; this animal had conical, pointed teeth, lived in forests and was solitary. Around 50 million years ago, rainfall decreased, temperatures decreased and the forests receded to grasslands. The next ancestor of the horse now had abundant grass to eat rather than lush forest vegetation, but grass is difficult to digest because the main component is cellulose. Bacteria, fungi and other microorganisms can break down cellulose and developed a mutually beneficial relationship with horses – they gained a place to live (in the horse's intestine) and the horse benefitted by now being able to access broken down cellulose. The ancestral horse thus adapted from forest to grass and this necessitated another change: eating grass means you also ingest silica from the soil and silica is very hard and grinding it wears teeth down. To deal with this, horses evolved hypsodont dentition – teeth that continue to erupt throughout the animal's life, as opposed to human dentition, brachyodont dentition, where our adult teeth erupt once. We discuss the effects of this dentition in relation to ageing in Chapter 5. Next in the evolutionary line was Merrychippus – around 35 million years ago. Merrychippus represented the first evolutionary progenitor of the horse with an enlarged hindgut and the first hypsodont dentition. This ancestor of the horse was more horse-like than its predecessors and probably lived in herds. Whereas living in a herd in a forest might be a disadvantage, it is an advantage in an open area where camouflage is not as effective. So, this is the background of the evolution of our horse's digestive system; we must now consider how horses spend their time, which will help us in understanding how nutrition and behaviour are linked.

Feral horses in the Carmargue region of France spend 60 per cent of their time eating, 20 per cent standing, 10 per cent lying down and 10 per cent doing other activities such as playing and mutually grooming (McGreevy, 2013). Horses with access to grazing will do so for 15–18 hours a day, trickle feeding on grass. If we compare this with a typical domestic horse (remember Swayze's day?), they are often fed 70 per cent of their diet as concentrated palatable feed and 30 per cent forage and their time budget is likely to be 65 per cent standing, 15 per cent eating, 15 per cent lying down and 5 per cent doing other activities (McGreevy, 2013). Concentrate food is easy to chew and eaten very quickly; even the average haynet only lasts for approximately two hours. In summary, horses now often have a starch-rich, energy-dense diet and restricted ability to use the energy – as Dr Andrew Hemmings says, "Our feeding practices are out of date – 50 million years out of date".

To understand how starch, behaviour and the brain are linked, we must consider how the horse digests food. Food arrives in the stomach in a lump and is passed to the small intestine where the digestion of starch begins. The presence of starch in the small intestine leads to glucose liberation; the timing of this release differs between food sources: release from oats peaks two hours post feed whereas release from alfalfa is much more gradual. Some starch is broken down in the first part of the small intestine and if there is too much starch to be broken down here, it continues to the large intestine where it can cause problems. Bacteria in the large intestine have evolved for breaking down grass but the breakdown of starch leads to an increase in acidity and some bacteria die. Indeed, studies comparing the reactivity of horses fed with low- and high-starch diets (Hale *et al.*, 2011; fed for four weeks with each diet) showed a significant reduction in reactivity/hypervigilance with a low-starch diet.

To fully understand the link between what we feed our horses and their behaviour, we need to consider a structure in the brain called the amygdala. The amygdala deals with the negative aspects of emotion such as anxiety, fear and aggression. In early studies using live animals, when this area was activated primates flew into uncontrollable rage and fear aggression. There has been little work on behavioural neuroscience in horses but studies comparing the amygdala of different species have found that those with higher fear responses have a larger amygdala. Zebras have larger amygdalae than horses do. We do not yet know if there is any correlation between breed of horse and size of amygdala but this work provides yet more evidence that feeding should be tailored to individuals – feeding a high-starch diet to an animal with a large amygdala is likely to be more harmful than to an animal with a smaller amygdala.

Two other regions of the brain are also important in considering the link between feed and behaviour: the hippocampus and the septal nucleus. Both are involved in the generation of pleasure and are commonly considered to be the reward centres of the brain, especially the septal nucleus. The hippocampus also has a role in short-term memory and spatial navigation. It is the hippocampus that is the primary area of the brain affected by feeding, especially by starch, which yields glucose.

To understand why feeding starch to horses can make them more "excitable", we need to consider neurotransmitters, chemicals in the brain that relay messages. One such neurotransmitter is serotonin, otherwise known as the "happy hormone", particularly known for its role in the action of chocolate and some antidepressants. The drug MDMA (ecstasy) causes a huge increase in hippocampal serotonin leading to a behavioural change – people taking this recreational drug become hyperactive, often able to dance for 24 hours! A small rush of serotonin, however, can be calming, hence being part of the mechanism of a common antidepressant – for optimal effect, we need a manageable elevation in availability of serotonin and not rushes.

Increased starch in horse feed, leads to increased glucose, which results in increased hippocampal serotonin, which results in excitable behaviour. This could be followed by a downturn and depression, especially because the horse is even less prepared for glucose than humans so is likely to experience more of a "crash". Another factor affecting serotonin is ambient light. In low light, from shorter days, there is also a reduction in serotonin. So consider Swayze in his stable all day – his serotonin turnover will be affected by the light available as well as his feed. In addition, the pathway that converts tryptophan to 5-hydroxytryptophan to serotonin to melatonin is affected by feeding. The enzyme that converts tryptophan to 5-hydroxytryptophan is called tryptophan hydroxylase and is encoded by a gene. In humans, individuals who have a mutation in this gene produce a less effective enzyme, resulting in reduced serotonin and are more prone to depression (along with environmental factors). This is also likely to be the case in horses – some individuals are likely to be more prone to depression than others.

The most plausible link between glucose release and the level of serotonin in the brain is that when there is more starch around, more tryptophan crosses the blood–brain barrier and is converted to serotonin. In turn, serotonin can be converted to melatonin, which is responsible for helping us switch off and go to sleep and if this happens, there is a danger of low serotonin, which affects sleep. You can now purchase lights to shine into horses' eyes to hasten the arrival of breeding. A behaviourally minded approach would be to provide the mares with access to sunlight.

We can manage the potential problems of a high-starch intake by using slow-release feeding systems, using certain supplements or changing the order of feeding. Some supplements can help to reduce the release of glucose and others can act as calmers to "cover up" the response. There is the opinion that calmers work on owners rather than horses! One view is that supplements work if the system needs rebalancing – for example, if a horse is deficient in magnesium then a magnesium supplement might work but will otherwise have little effect. A behaviourally minded approach would again point to the importance of keeping a horse at grass and minimising concentrated feeds.

So, the two key messages with respect to nutrition and behaviour are that first we must consider the evolutionary origin of the horse – avoid feeding a diet that is 50 million years out of date. Second, although we can mitigate the effects of too much starch in the diet, we must recognise that most behaviour problems could be avoided by addressing the cause – we should be feeding forage and not concentrate.

Stereotypical behaviour and confinement

In some cases of chronic stress (which we will cover in Chapter 8), animals start to develop abnormal behaviours as coping strategies and perhaps the most common abnormal behaviours we see in equines are stereotypies. Stereotypical behaviour has been defined as "a repetitive, invariant behaviour pattern with no obvious goal or function" (Mason, 1991). These behaviours are not observed in animals in the wild and are considered abnormal. The most common equine "stereotypies" are cribbing and windsucking, weaving and box-walking. However, there are other horse stereotypies including repetitive head nodding, tongue playing/lolling and lip smacking. Stereotypies are usually performed when the horse is aroused, for example, waiting for a feed, which they know is coming soon. Sadly, stereotypical behaviours are often misinterpreted as quirks or "funny" behaviours the horse does to get attention – missing the underlying cause, which is management (or historical management) that causes stress; the stereotypical behaviours are an attempt to cope with that stress.

Research has shown that the incidence of stereotypies in the domestic horse is concerningly high. Studies have shown that as many as 32 per cent of dressage horses may perform an abnormal behaviour such as stereotypy (McGreevy et al., 1995). Studies have also shown that management types and training/exercise regimes affect the incidence of stereotypy with racehorses and dressage horses having the highest incidence, and leisure horses having the lowest.

Scientist Kathryn Houpt estimated that horses chew between 10,000 and 40,000 times a day (lower range were fed predominantly pellets, higher range predominantly hay and grass; reviewed in Houpt, 2011). The horses observed to chew the most lived outside full-time while the ones with the lowest chew rates were stabled. She made an observation in her study that she has never seen a horse crib when in an environment that allows the horse to chew towards the higher rate. If chewing is prevented through confinement or management techniques that prevent it, the horse seems to fulfil the need in another way.

One study found that combining ad-lib forage with provision of a horse lick helped to change the behaviour of crib-biting horses (Moore-Colyer et al., 2016). Crib-biting horses changed their behaviour when they had access to a lick 40 times more during the 30-minute observation period compared with non-crib-biting horses (they changed between eating hay, licking the horse lick, looking over the door, drinking, crib-biting, etc.). Although the frequency of crib-biting did not significantly alter when the horse had access to the lick, the crib-biters licked the supplementary lick 1.5 times more than the non-crib-biters, giving them another activity to engage in. Thus, abnormal behaviours are thought to be the result of horses trying to cope with prolonged confinement. It is thought that mouth-based abnormal behaviours, such as windsucking and crib-biting, develop as a way to meet the need to chew for 15–18 hours a day and movement-based abnormal behaviours, such as weaving and box walking, are the horse's attempt to meet the natural walk-and-eat behaviour.

There are some medieval-looking gadgets on the market to prevent cribbing including straps (which cause pain when cribbing is attempted), shock collars (deliver an electric shock when the horse tries to crib thus stopping the behaviour) and adding metal rings on the teeth (causing pain when the horse tries to crib). Surgery (removing the muscles required to crib) is also sometimes undertaken. Weaving grids prevent weaving, and the traditional way of dealing with box-walking horses is to add a bale of straw to the stable, although this often induces jumping. Every horse equipment catalogue now has a section of stable toys, reflecting a growing understanding that horses benefit from "things to do" during confinement. The problem with all these gadgets is that they focus on the symptom, not the cause. Prevention of stereotypical behaviour is a welfare concern because animals performing stereotypical behaviours are doing so as a coping mechanism and if we remove that, without addressing the cause, we remove their ability to cope.

It is important to also understand a little of the neurobiology associated with stereotypical behaviour. The brain structures involved are the nucleus accumbens (the pleasure centre in the brain), the ventral tegmental area (the "dopamine factory") and the substantia nigra (which also produces dopamine). When rodents are given cocaine, dopamine increases and they are more awake, alert with faster reactions – all good results when there is a threat of predators. Dopamine is not only a happy hormone but also a stress hormone depending on which brain region and which receptor it is binding to. In the nucleus accumbens, the dopamine receptors change with prolonged use and the structure itself changes with repeated exposure. There is some evidence that crib-biting horses have more receptors for dopamine than non-crib-biters, suggesting that horses are getting pleasure from cribbing and the behaviour represents a coping mechanism with behavioural restriction such as confinement. If we prevent crib-biting the result is increased illness due to the unaddressed stress and removal of the coping mechanism and decreased performance. Incidentally, a good way to tell if a horse is a cribber is to produce a highly palatable reward. A habitual cribber will crib after ingestion of the reward – starch-rich palatable feeds induce cribbing. Due to the physiology of stereotypical behaviours, it can be difficult to address them completely through changes in management; such behaviours can be considered "behavioural scars", scars that remain after the cause has gone.

Meet Jane Myers and The Equicentral System

One of the most pioneering people in changing the management of modern-day horses is Jane Myers. Jane is a bouncy, passionate advocate for horses now but her journey into studying horses academically was not easy. Jane was born with significant hearing loss and went through school without hearing aids. Throughout school she sat at the back of the class and daydreamed about horses. She left school at 16 without a single exam pass result and started her first job with horses. After several years, she decided that she needed some qualifications, eventually obtaining a master's degree in equine science. Following her MSc, she undertook a period of research on the grazing behaviour of horses and is the author of many books (e.g. Myers 2005).

Now, Jane and her husband Stuart, provide one-day workshops about sustainable horse property management called "Healthy Land, Healthy Horses". They also provide workshops and training on horses in the community, the environmental aspects of horse ownership and understanding horse owners for natural resource managers (such as land care officers in Australia, water catchment organisation employees and local council workers). They provide a consultation service for horse property owners on horse property design and layout.

Jane and Stuart set up Equiculture and through this evolved The Equicentral System – a set of principles that is used to design a set of fields and shelters for optimal grass and environment management. Based on understanding horse behaviour, the system is beneficial for horses and for the environment and thus is the best example of behaviourally minded horse keeping I can think of. It soon became clear to them that once you start to do the right thing for the horse and for the land, it works well for the environment too, a win–win scenario.

In a nutshell, there are four main elements to The Equicentral System:

1. Providing a yard area of hardstanding where hay is available *ad lib* – this enables time off the grass, still able to be sociable and with the opportunity to eat low-energy fibre.
2. Rotating paddocks so some are rested while others are used. Can combine with strip grazing if needed. Stuart has a rule of the "stubby" (Australian term for beer bottle) – if the grass is the height of the beer bottle on its side, then it is time to rest it. If it is the height of the beer bottle standing up, time to graze it.
3. Arranging gateways and pathways so that the horses can come to the hardstanding on their own helps to avoid poached and muddy gateways.
4. Think of being a grass farmer rather than a horse owner!

Like many other equine professionals, Stuart and Jane teach that we cannot (should not) look at one aspect of horse management in isolation, nor focus on one part of the horse such as the hooves or gut; if we do, we lose sight of the bigger picture. Jane and Stuart introduce ideas around behaviour, nutrition and welfare, and weave that into environmental management and knowledge about grasses. Jane gives an example of how horse behaviour and management are linked:

> The 'hanging around' behaviour that horses carry out in gateways is a behaviour that has been created by humans, it does not happen in the wild. The only reason that horses stand for hours at a gateway is because the gate is shut and it is preventing them from moving to where they want to be. This behaviour causes problems including skin problems, muddy or dusty gateways and fence injuries. By making a simple change to The Equicentral System, which provides different areas for different behaviours, this hanging around is used to our advantage, whilst at the same time removing the problem and not stressing the horse in any way. The horse is allowed to move to where they want to be (to where the shade/shelter/water etc. is situated), mud/dust is avoided as are all of its resultant problems.

Jane's studies found that a grazing bout lasts up to three hours and stops when the fibre volume indicates that the stomach is full. Horses only move to get to food; no grazing means no movement. Once this part of equine behaviour is understood, people can apply it to what is needed for good management. While horses are grazing, they are out on the pasture, but when not grazing they should be able to stand in a surfaced area that is designed for such behaviour. Horses do not want to stand in mud (not for long periods anyway), they do it because they have no choice and The Equicentral System gives them choice.

Horses are "pre-programmed" to eat when food is plentiful, to put on weight to survive winter or the next drought. However, if you restrict a horse's diet, his body tells him that there must be a food shortage and therefore he must eat as much as is available. Not letting this "starvation mode" behaviour become the norm vastly improves the health of your horse. Jane and Stuart are convinced that most domestic horses have developed man-made "eating disorders".

The approach has great take-up, thanks not only to the fact that it works but also to Jane and Stuart's passion for what they do. Even people with considerable experience in managing horses on land are changing. Jane gives an example:

> We had one couple in their seventies who have been managing their highly respected stud along traditional lines for years, but it was becoming too hard to manage. By making a few simple changes over to an Equicentral System, their management has become much easier and they have reported that their horses are much calmer and easier to handle than they were before.

Swayze: six months later

At the beginning of this chapter, we met Swayze, who was living a life in confinement. He belonged to a client of mine and all the horses were kept the same way on that yard of about 60. He was becoming increasingly aggressive in the stable and difficult to handle. He would turn his back on people and try to kick them when they entered his stable and was becoming increasingly less tolerant of being brushed and having his rugs done. As a behaviourist, I did not merely treat the issues as training needs, I wanted to address his environment to better meet his needs. However, there was the usual problem of a busy yard, a high turnover of staff and a very robust culture of keeping horses that way – abnormal behaviour was the norm in this situation.

The owner understood that, ideally, he would live out a lot more, socialise with other horses, and so on but was not prepared to make those changes for various reasons. The yard was open to changing the turnout so that the same horses were neighbours each day. Over time, the horses were showing a little interest in each other and even attempting pro-social behaviours such as mutual grooming bouts. Due to some changes on the yard, a U-shaped stable block became available and we arranged for Swayze and his field neighbours to move there. There was a fence across the wide end of the U and we were able to add temporary fencing across the middle of the U (parallel to the wide end) so that pens were created outside the stables. This provided the horses with choice in a way that they had not had since moving to that yard; for some of them, this was more choice over their environment than they had ever had. Food was always given inside the stables to avoid any resource guarding behaviour across the temporary fences. The horses had the choice of being in or out of their stable and most of them developed the habit of grabbing a mouthful of hay and walking out into the pens to eat it socially with the others.

The stables and pens were all enriched using many of the ideas listed in this chapter. Different horses had different preferences for the various types of enrichment and a new culture was developing in the U-shaped yard. Owners were becoming increasingly interested in the lives of their horses when the owners weren't present and they started to leave observational notes to each other. To muck out his stable, staff at first shut Swayze in the pen when they cleaned his stable and vice versa but in a short time Swayze was overlooking proceedings in a friendly way and the grooms felt safe without the barriers. Together with coaching the owner through gentle handling techniques, he learnt to accept being groomed again and it wasn't long before he actively enjoyed it. We did this loose at first in the pen and stable area so that he could move away if he wanted to. Again, adding autonomy into Swayze's life was a key part of changing his behaviour.

Final thoughts

Perhaps the key to having behaviour in mind when keeping horses is their actual management and care. This chapter clearly shows that the best way to meet the needs of horses is for them to live out as much as possible, with access to ad lib forage and other horses for social interaction. If horses need to be stabled, both temporarily or as part of a longer-term management system, then with a little thought we can use enrichment to go some way to trying to partially meet their needs.

References

Bulens, A., Dams, A., Van Beirendonck, S., Van Thielen, J. and Driessen, B. (2015) A preliminary study on the long-term interest of horses in ropes and Jolly Balls. *Journal of Veterinary Behavior: Clinical Applications and Research*, 10(1): 83–86.

Glunk, E.C., Hathaway, M.R., Weber, W.J., Sheaffer, C.C. and Martinson, K.L. (2014) The effect of hay net design on rate of forage consumption when feeding adult horses. *Journal of Equine Veterinary Science*, 34(8): 986–99.

Goodwin, D., Davidson, H.P.B. and Harris, P. (2002) Foraging enrichment for stabled horses: effects on behaviour and selection. *Equine Veterinary Journal*, 34(7): 686–691.

Goodwin, D., Davidson, H.B.F., and Harris, P. (2005) Selection and acceptance of flavours in concentrate diets for stabled horses. *Applied Animal Behaviour Science*, 95(3–4): 149–164.

Gumm, G.G. (1991) Physiology of the Eye. In Gelatt, K.N. (Ed.), *Veterinary Ophthalmology*. Lea & Febiger, Philadelphia, PA, USA. pp. 124–161.

Hale, C.E., Hemmings, A. and Bee, S. (2011) The effects of a high starch, cereal-based diet compared to a low starch, fibre-based diet on reactivity in horses. In *Applied Equine Nutrition and Training*, Wageningen Academic Publishers, Wageningen, The Netherlands. pp. 227–231. DOI: 10.3920/978-90-8686-740-0_17.

Houpt, K.A. (2011) *Domestic Animal Behavior for Veterinarians and Animal Scientists*. Wiley-Blackwell, New Jersey, USA.

Jørgensen, G.H.M., Liestøl, S.H.O. and Bøe, K.E. (2011) Effects of enrichment items on activity and social interactions in domestic horses (*Equus caballus*). *Applied Animal Behaviour Science*, 129(2): 100–110.

Kwiatkowska-Stenzel, A., Sowińska, J. and Witkowska, D. (2016) The effect of different bedding materials used in stable on horses behavior. *Journal of Equine Veterinary Science*, 42: 57–66.

Lamoot, I. et al. (2004) Eliminative behaviour of free-ranging horses: do they show latrine behaviour or do they defecate where they graze? *Applied Animal Behaviour Science*, 86(1): 105–121.

Mason, G.J. (1991) Stereotypies: a critical review. *Animal Behaviour*, 41: 1015–1037.

McAfee, L.M., Mills D.S. and Cooper, J.J. (2002) The use of mirrors for the control of stereotypic weaving behaviour in the stabled horse. *Applied Animal Behaviour Science*, 78: 159–173.

McGreevy, P. (2013) *Equine Behaviour: A guide for veterinarians and equine scientists*. 2nd edn. Elsevier, London.

McGreevy. P.D., French. N.P. and Nicol. C.J. (1995) The prevalence of abnormal behaviours in dressage, eventing and endurance horses in relation to stabling. *Veterinary Record*, 137: 36–37.

Mills, D. S., and Davenport, K. (2002) The effect of a neighbouring conspecific versus the use of a mirror for the control of stereotypic weaving behaviour in the stabled horse. *Animal Science*, 74(1): 95–101.

Mills, D. and McDonnell, S. (2005) *The Domestic Horse: The origins, development and management of its behaviour*. Cambridge University Press, Cambridge.

Moore-Colyer, M.J.S., Hemmings, A. and Hewer, N. (2016) A preliminary investigation into the effect of ad libitum or restricted hay with or without Horslyx on the intake and switching behaviour of normal and crib biting horses. *Livestock Science*, 186: 59–62.

Myers, J. (2005) *Managing Horses on Small Properties*. Landlinks Press, Collingwood, Australia.

Ödberg, F.O. and Francis-Smith, K. (1977) Studies on the formation of ungrazed eliminative areas in fields used by horses. *Applied Animal Ethology*, 3(1): 27–34.

Yarnell, K., Hall, C., Royle, C. and Walker, S.L. (2015) Domesticated horses differ in their behavioural and physiological responses to isolated and group housing. *Physiology & Behavior*, 143: 51–57.

Chapter 2
Breeding
Jenni Nellist

In this chapter, I will consider how the natural behaviour of horses surrounding reproduction interfaces with modern stud management and how the breeder can manage their horses with behaviour in mind. In particular, I will introduce the implications of the effect of living environment and experiences on the transfer of genetic information, the effect of stallion and mare management on their general welfare, ability to conceive and carry a healthy foal to term, as well as on courtship behaviour and the events surrounding covering. I will then turn my attention to the foal and their behavioural development and management through to weaning and puberty. Throughout the chapter I will discuss effective, behaviourally minded practices, including example case studies. The aim of this chapter is to promote positive equine welfare, fit for the twenty-first century, through responsible breeding with behaviour in mind.

A stallion called Buck

Buck is a young stallion from a strong Western Sport pedigree. His owner, Dave, has high hopes of selling youngstock to Western riding enthusiasts, given that other horses in Buck's family have calm, trainable temperaments and have excelled in various Western disciplines and all-round ranch work. Buck is yet to be broken to ride but, at the age of 4, he's already had one season breeding mares, and his foals appear to be calm and biddable. Buck is kept stabled all winter in a box stall where he cannot touch or see another horse unless he puts his head over the door – and even then, there is an empty box between him and the other stallion kept on the stud. He is given a few hours turnout on a timeshare basis with the other stallion to prevent the two from coming into physical contact and potentially fighting. In the spring and early summer, he is used for covering mares and during that time he is still predominantly stabled with only short periods of turnout – still time sharing with the other stallion. On the days he is used for covering mares, Buck's groom puts a head collar on him, with a lead rope trailing, then leaves Buck in his stable. A mare is then presented just outside his box to tease her and check for optimum timing of breeding. If she is deemed ready, she will be led forwards into the yard outside and Buck's groom will pick up his lead rope and open the stable door to bring him up to serve the mare.

The closely managed scenario described above illustrates some common practices in horse breeding. However, it is not only the act of keeping horses and managing their covering that can be so highly controlled but also procedures around the birth, the management, handling and weaning of foals, and the management and handling of weanlings as they become adolescents. How do horses breed and raise their young in natural, free-living populations? What are the effects of controlling horse breeding and how can we optimise it in ways that allow for more natural reproductive behaviour in horses? This chapter will explore those issues and will introduce ideas and practices from passionate and thoughtful horse breeders keen to share their positive experiences of breeding and producing healthy and "sane" young horses.

Runs in the family

There is nowhere more pertinent for discussing the effects of nature versus nurture than in breeding animals. On the one hand, the horse's future characteristics can be put down to his genetic inheritance, and on the other, these characteristics are determined by how he is nurtured and raised. Selecting good parents, sound of body and mind, with desirable bloodlines to increase the probability of breeding to type is universally accepted good practice. What is less commonly known is that the environment of the forebears also affects the horse we are bringing into the world, by altering how the inherited genes are expressed; this is called "epigenetics".

Epigenetics enables certain genes to be "switched on" or "switched off", without altering the underlying DNA itself. These changes happen in response to the environmental challenges faced by the horse, at any point throughout their lifetime. What is really important to horse breeding is that epigenetic changes can be transferred to the next generation through both parents. In humans, a mother's exposure to stress, such as to mothers during the Dutch famine of 1944–1945, or a father's drug and alcohol habits can affect their future children. The children of the Dutch mothers were more prone to heart disease and type 2 diabetes (Painter et al., 2005), and in mice, fathers exposed to alcohol or cocaine had more anxious offspring compared with unaffected fathers (Powledge, 2011). On a more positive note, mice living in rich, social environments where they could communally rear young had offspring that were calmer and bolder than those of mouse parents kept isolated in barren cages (Powledge, 2011). These effects normally last one to two generations, but can last as long as five generations. Figure 2.1 shows a bonded semi-feral mare and gelding mutually grooming; mares will gravitate towards living with a male horse, entire or gelded, for both companionship, a male parent who will protect and play with their offspring, and for increased feelings of security.

Epigenetic research into horses is scant; investigation into the "methylation" of DNA is one means of understanding the process. One male Thoroughbred horse has been compared with a Jeju, a native Korean pony, also male and of similar age. Differences in the expression of genes controlling muscle development demonstrated evidence of their different epigenetics (Lee et al., 2014). It is thought such information might have a role in selecting for better racing ability in the Thoroughbred but it is early days for wider applications in horse breeding. However, given what we know about the epigenetic effects of the mammalian parents' environment on the future health and behaviour of the offspring, such a seemingly irrelevant finding nevertheless demonstrates

Figure 2.1: Cefn Hugo Boss Femme grooms with her surrogate male consort, the gelding, Cefn Busy Chap. Photo: Jenni Nellist.

epigenetic effects in horses and underlines the fact there is still much to learn about them. To breed wisely would include being mindful of the parent horses' current and previous life experiences, and their possible effects on the temperaments and coping abilities of the offspring. Some veterinary professionals are examining the effects of the environment on the prenatal mare and early foal development in relation to disease and its prevention (Dindot and Cohen, 2013).

> **Given what we know about the effects of genetic make-up and epigenetics on the next generation, the following are key points to adopt into practice:**

- Select friendly, resilient horses who have successfully coped with moderate stress, such as successful riding horses, or others that thrive in the types of environment you will expect the offspring to live in later in life.
- Keep broodmares in compatible family groups, from before conception and right through to weaning and beyond.
- Wean youngsters with compatible others.
- Enable stallions to live in compatible social groups: bachelor groups with colts, well away from mares (this keeps the testosterone and aggression levels down). Alternatively, keep them with their mares.
- Avoid exposing horses to extreme levels of distress and chronic stress – in management, handling and training.
- Apply learning theory and use the LIMA approach to handling and training (LIMA stands for least invasive, minimally aversive).
- Provide a fibre-based diet, using oils to supplement for additional energy requirements.
- Even overweight broodmares should not be starved or subject to a fibre intake lower than 1.5 per cent of their own body weight.

> **Case study**
>
> Colin Thomas is a stalwart of the Hill Ponies of Wales, the collective organisation representing all the Welsh Pony and Cob Society affiliated Hill Pony Improvement societies. Despite the Welsh Mountain pony being popular worldwide, the semi-feral Welsh Mountain ponies and the bloodlines that contribute to their success on the hills of Wales are in decline. In order to preserve the resilience of the hill pony within the breed, Colin runs his registered broodmares on the hill all year round apart from an annual gather. His youngsters are particularly even-tempered and integrate well on the hill when they are turned back out as 2-year-olds. They have proven to be equally successful as show and children's riding ponies.
>
>> I breed semi-feral ponies that graze 1500 feet above sea level – the mares that I keep have all been bred on the Dowlais hill, the mares must have specific characteristics that retain the hardiness to survive very harsh conditions of the environment that they live out on throughout the year. Foals that are surplus are sold or kept on lowland; their versatility and excellent temperament go on to make them ideal leading rein or performance animals. The stallions we use must have breeding lines and characteristics that will pass to the offspring.
>
> It's not just hill ponies who do well to live in family groups, horses intended to be companion animals, particularly family horses, are best bred from horses already proven and nurtured in this role. Jackie Hughes bred from her ex-racehorse to produce another riding horse for herself. Her mare was a resilient and friendly character who enjoyed speed. Jackie found a friendly PRE stallion (Pura Raza Española or Spanish stallion) with a good, trainable temperament to be the father. To save her mare the stress of transportation to stud, she chose to use artificial insemination (AI), allowing her mare to remain within her stable social group in which she had lived for many years prior to being bred from. When I asked Jackie what she valued in the parents of her young horse, she told me: "Good conformation, no congenital health problems, ease of handling and people orientated".

Stallions

It is now established that individual housing of horses, especially when they cannot easily see or touch other horses, is chronically stressful for them (Yarnell *et al.*, 2015). And yet stallions are often kept in this way, often leading to some antisocial behaviour and/or being too enthusiastic and even aggressive to mares when brought out to cover them. This is often largely because when managed in such a restrictive way, the stallions have frustrated behavioural goals – they have pent up fears and frustrations because they are not able to interact normally with other horses. Ethologists call this "rebound" behaviour – when an animal released from its barriers, performs certain behaviour with a lot of energy and for longer than the normal length of time – the saying "like a kid in a sweet shop" springs to mind.

Sadly, the abnormal behaviour of self-mutilation is relatively commonly encountered in breeding stallions – sometimes biting their own skin, potentially causing significant wounds. As with many behaviours, as we will discuss in great detail in Chapter 8, the cause could be pain or disease and should be investigated by a vet. However, if there is no medical reason for the behaviour, the cause might be the restricted management and associated behavioural frustration stallions are often subjected to, perhaps also the performance of mating behaviours without the natural courting elements of the equine ethogram. Self-mutilation behaviour can develop into slowed down, repetitive "stereotypical" behaviour (McDonnell, 2008), widely understood to be a coping mechanism to deal with behaviourally challenging situations.

In the natural world of free-living horses, stallions live mostly in social groups, either with a band of mares, with a band of bachelors or in a mixed herd. A stallion in this situation is able to communicate with other horses all the time, apart from when he is asleep, practising his sparring skills in play with other stallions, and tending and watching his mares and foals once he has successfully become a harem stallion and father. In these situations, there is no managed breeding, and no thwarted motivation to interact or to breed. I have spent many years following free-living horses local to my home in Wales, UK, studying their behaviour for my honours degree, leading field visits, and photographing and filming them. I have watched stallions from different harem groups meet up, not to fight, but to play. I have observed bachelor groups of colts, minding their own business, leaving the other groups alone while they eat, sleep and play amongst themselves. When they do get the opportunity to breed, they will first perform courtship behaviours, interacting with their mares, not getting too amorous when the mares are not ready to breed, and being approached by them most frequently to mate when they are in season.

Bachelor stallions, those that live only with other males, tend to have lower testosterone levels than harem stallions who are actively tending their mares. Likewise, where more than one stallion resides in a feral horse group, the beta, or lieutenant stallion, also tends to have lower testosterone levels, because he does not have breeding rights. This means that keeping stallions together has the potential to lower some or all the stallions' testosterone levels and reduce their sex drive, even to nothing if they are also fearful of painful events surrounding the covering process. Finding the practical solution to preventing a stallion from becoming socially isolated therefore depends upon his status in relation to other stallions in the breeding barn, and on his access to mares aside from the covering shed.

Some points to consider with keeping stallions:

- Avoid keeping them alone, especially in single box stalls where the only view is over the box door.
- If being kept in a stallion barn reduces libido in a (typically) lower status male, consider keeping him with at least one mare.
- Outside of the breeding season, keep stallions in social groups, ideally at pasture for at least part of the day – either with other stallions, or with mares.

To introduce a stallion to pasture breeding, or turnout with other stallions, introduction should be made gradually, if possible, as described in Chapter 7. If introduced in a more traditional way of turning the animals out together, expected initial interactions might involve running around. It is therefore important that the animals are familiar with the area, which decreases stress and reduces the risk of injury. To turn stallions out together, it is best to do so after stabling them next to each other, to create bachelor group status and lower testosterone. Also, do this outside of the breeding season, to further ensure testosterone levels are low as they can be, and to be safe, turn them

Figure 2.2: After a brief encounter over some bread, rival stallions agree to part company. Photo: Jenni Nellist.

out well away from mares, such as was accomplished by the Swiss National Stud with their carriage-driving stallions, and is normal procedure at their establishment (Freymond *et al.*, 2013). Their stallions were at first aggressive to each other, mainly ritualised posturing rather than overt aggression such as bites and kicks making contact, but this antagonistic behaviour diminished within four hours and after three months, a stable dominance hierarchy emerged. Figure 2.2 illustrates how familiar stallions use ritualised posturing to avoid overt aggression. The stallion in the foreground is about to trot away from the other, who has claimed his harem's right to some waste food left out on the hill by humans.

Making a foal

Let's talk about sex. This is what equine courtship behaviour is all about – there is a biological advantage in each partner communicating well in order to be at ease during mating and maximise the possibility of conception.

In most free-ranging conditions, feral horses breed extremely well, and indeed abandonment of horses and ponies on common land in the UK during the "horse crisis" of the late 2000s and early 2010s has led to high numbers of animals requiring to be removed – they have been too successful at breeding. Where there is very little or no management, mares usually live with their stallion all year round. Indeed, there is some evidence he is *their* stallion; mares can be "stolen" by another stallion only to return to their previous group, move to another stallion's harem if they prefer him, and even actively reject stallions they do not want – usually immature ones (Boyd *et al.*, 2016). There was also an incident where another hill pony breeder ran two stallions; he found some mares leaving their usual companions to run with the other, preferred stallion. I've also observed that

> **Case study**
>
> Mary Prewitt breeds Connemara ponies and strives to do so with behaviour in mind. Her young stallions live out their lives in company; once they have been weaned they are kept with other colts and geldings, enabling them to continue to practise their social skills. Sometimes a colt is kept with a mare already in foal, generally in the winter when the stallion's sex drive is lower, but still making sure to keep a close eye on them to monitor interactions. Mary says:
>
>> I cannot emphasize how important proper socialisation is for the young stallion who is going to live naturally with his mares and foals. Taking a stallion who has always lived in isolation and turning him out with mares would be disastrous for all involved.
>
> When the mare is in season, Mary hand breeds her to the young stallion and immediately turns them out together in the pasture. She says, "I prefer to use an experienced mare the first time and ensure the mare is "very hot" in heat". Mary has seen a young stallion cover a mare 15 times in the first hour! By the time the mare is coming out of season, the young stallion has had the edge taken off his ardour, and he's more willing to accept a "no" when she starts declining his advances.
>
> Mary continues:
>
>> I also like to have another mare who's coming into heat at this time and repeat the process of hand breeding and then turning them out together. As stallions get older and more experienced, they generally learn how to tease a mare properly and not get too pushy before she's ready. But while they are young, I monitor and observe to be sure that the stallion is learning proper and polite breeding behaviour.

when mares outnumber stallions, there is little to no fighting between resident stallions; they even meet up for play dates! The mares are happy with their mate and he doesn't need to search elsewhere. They are also very tolerant of young stallions who come to flirt and mate with their daughters, as has also been reported elsewhere (Boyd *et al.*, 2016). In free-ranging conditions, both sexes are able to make choices about who they breed with.

Natural tending and courtship involves the stallion checking the scent of his mares as he goes about his normal daily life of grazing, resting, playing with youngsters and keeping the group together. Within days before foaling and immediately after, the chemicals released by the mare pique the stallion's interest and he may make sexual advances towards her. She will reject these by kicking out at him. If he heeds these messages all will be well but to avoid such harassment, many breeders choose to separate their mares for foaling to avoid accidents and preventable stress. The next time that the mare will smell interesting is during her "foal heat". The stallion can be seen performing the flehmen response, raising his head and curling his top lip to draw the scents into the Jacobson's (or "vomeronasal") organ to further analyse the mare's sexual and social pheromones. This in return will increase his sexual ability, and he'll also become more alert to possible threats from other stallions, and even people who may wish to catch a mare and lead her away. Under such threat, a stallion will position himself between the threat and his mares, square up to it, squeal and stamp his foot ("mine!") before taking any necessary further action. All good stud hands know to catch the stallion first before going to attend to any mares! Another part of the stallion's defence of his mating rights are to cover his mare's urine and faeces with his own urine, having the effect of breaking up the compounds advertising her receptive, keeping this information to himself (Kimura, 2001).

The stallion's sexual advances may be triggered by the mare's scent or her behaviour – urination, approaching him and squatting in oestrus stance. However, in more than 88 per cent of observed feral horse matings, it is the mare who has approached and solicited the stallion, encouraging him to join the courtship (McDonnell, 2000). Her classic squatting and lifting her tail to one side, enabling the stallion to both see her vulva winking, better smell her sexual pheromones, and prevent her tail hair becoming snagged in the process, send him a very clear message that she is receptive and safe to approach. With the reassuring confidence that she is safe to approach, the stallion will approach her, neck arched and telescoped forwards, often give a deep nicker, having the effect of increasing her enthusiasm for covering. Provided she doesn't then move away or squeal and kick out, he will touch, nibble, nip and bite at her flank before testing her response to putting his head over her neck before finally mounting. He may even mount several times, since she may still not be completely receptive. Finally, with the mare squatting to take his weight and position herself, the stallion gains intromission and they copulate. Ejaculation happens at around 20–25 seconds (McDonnell, 2005) and is simultaneous with the stallion flagging his tail. Afterwards, there is a brief rest in which the stallion's engorged penis shrinks and is comfortable to withdraw, then when the mare is ready, she steps forwards enabling the stallion to gently dismount. This sequence will happen many times per day when the mare is on heat, lasting two or three days.

When mares and stallions become difficult to manage for mating, it is because one or more man-made restrictions provoke it. In managed breeding, the mare and stallion are chosen for each other, based on the type of horse the breeder wishes to produce. It's not intended as a love match and so sometimes the mare is not interested in the mate chosen for her, particularly when she is teased by one stallion and is anticipating mating with him, and then another stallion is brought into the breeding shed. Likewise, the stallion may be incapable of reading or responding to her body language, and lack of sufficient courtship causes her distress, interrupting the covering process. Other reasons for refusing to stand include pain, such as from a recently reversed Caslick's operation where the edges of her vulva are sore or the twitch; fear, from being restrained through the bridle, being held in a corner and/or wearing hobbles. Many of these restraints are in conscientious effort not to harm a valuable stallion, but nevertheless prevent normal courtship with increased risk of negative results – a fearful mare who fails to conceive.

Such man-made intervention can cause the stallion problems too. He may be socially inept and struggle to read the mare's signals, get kicked as a consequence and become fearful of approaching mares, even leading to erectile dysfunction. Once having successfully covered a mare he may also have been forced to dismount too soon, causing pain to his still engorged penis and to his back as he is made to go backwards rather than simply down and forwards as when the mare walks forwards. Being so conflicted, between his sexual desire and anxiety about pain and rejection, he can become over aroused, even aggressive, or he may become shy and reluctant.

When hand breeding, it is good practice to allow as much courtship as possible. Use the same stallion to tease and cover the mare, and when she's receptive, bring her to the middle of some open space such as the yard where she's been teased. Allow the stallion to approach her from the side or rear and let him sniff her more if required. Keep handlers safe by having agreed escape routes – to turn each horse away from the other and to lead them in opposite directions, being sure that each horse is sufficiently trained to enable that. Once mating is over, allow the mare to step forwards when she is ready – this is more comfortable for all concerned and increases the probability of conception.

Pasture and paddock breeding both avoid the complications and restraints that can come with hand breeding and may be necessary to encourage a shy stallion to begin covering again, or even re-educate one who has become too vigorous in his approach – with careful risk assessment first of course.

In pasture breeding, a group of mares are turned out together at the start of the breeding season and the stallion is added to the group. This may be done instead of hand breeding or done after the stallion has performed other hand breeding duties to catch mares that might not have conceived earlier.

> **Fertility: pasture versus hand breeding**
>
> Welsh Mountain pony breeder, Colin Thomas estimates that he gets an 80 per cent live foal rate from his hill pony mares, who are free to mate as often as they like. This corroborates Sue McDonnell's findings that pregnancy rate after first oestrus in pasture breeding is typically higher than 75 per cent (McDonnell, 1992). Conversely, the historical results of hand breeding in the Thoroughbred industry have been lower, at 70 per cent pregnancy rate, but these days greater than 90 per cent is more usual thanks to a range of veterinary interventions (Allen and Wilsher, 2018). Historically, such pregnancy rates were due to high demand for quality stallions, making for only between one and four matings per oestrous, which might not have been optimally timed, copulation might have been sub-optimal and additionally there might have been venereal problems. A higher number of matings makes for more efficient copulations and for optimum timing for sperm to meet egg.

Mares and stallions in sport

With the advent of gradings for sports horses, it is now more common for both mares and stallions to be kept in work as much as possible whilst also producing the next generation of hoped-for sporting superstars. It's easier for stallions to be kept in competition-level fitness and to collect semen for artificial insemination than it is to remove them from competition for a strenuous covering season. Mares too, can conceive and have the embryo removed for transfer to a surrogate mare, and take only a limited time off work. For the mare to conceive and bring the foal to term, and then nurse to weaning, takes much longer, and she will need correct exercise to bring her body back to the strength required to compete successfully. This is often why it is elected to use embryo transfer (ET), or to retire a mare in her early to mid-teens to breed. Having more than one foal per year is also possible with ET. One of Charlotte Dujardin's rides, Mount St John VIP produced four foals by the age of six – three of them by ET (source: www.mountstjohneques trian.co.uk/).

Artificial insemination is not one-sided; an easy-to-handle mare might be a less stressed recipient than if inseminated by hand breeding

Photo: Suzanne Rogers.

but a stallion has to act as the sperm donor. This is relatively straightforward in a stallion familiar with hand breeding, who is led up to a receptive mare and instead of his penis gaining intromission to the mare's vagina, it is diverted to the artificial vagina, the interior lubricated and warmed to body temperature. One can only think this procedure might be somewhat frustrating for the mare who misses out, and so there is the option to train the stallion to use a dummy mare instead.

Jackie Hughes chose to breed her Thoroughbred mare to a PRE stallion. Rather than put her mare through the stress of travelling away from home and the rest of her stable herd of three, she chose to use artificial inseminatinon (AI). This necessitated timing the mare's heat so that she would be ovulating at the time of insemination. Jackie says, "By keeping an in-season diary, I could see that she was absolutely predictable, we only had to scan her to check a follicle had formed before the AI was done. Fresh semen was used – collected that day and she took first time".

Pregnancy

Once pregnant, the mare needs to be protected from significant sources of distress and chronic stress, but otherwise life needs to be as normal, rewarding and as meaningful as possible. To a horse mum this means living in a relatively stable herd in a safe environment. In feral herds, mares live in family units with their stallion, growing offspring, often other mares and occasionally other stallions. The stallion is key security against large predators, so it feels safe for a mare to live with a stallion. Semi-feral mares on common land in the UK will choose to live with a gelding year-round if there is not a stallion available. Families, even pseudo ones are important, for security and for positive emotional ties; the current recommended best practice is to keep pregnant mares outside in small, stable groups (Malschitzky et al., 2015).

Transport, intense and abnormal physical activity, colic, high parasite burden, premature and abrupt weaning of the previous year's foal and even experimental dosing of steroids, have all been shown to contribute to losing a pregnancy. The risk is higher earlier in pregnancy (the first 40 days) than it is at the end, so strenuous activity such as competing must stop, as should any teasing routine (Malschitzky et al., 2015). Later in pregnancy, when the conceptus has become the foetus, the foetus will respond to stress by increasing its own production of pregnancy-supporting hormones: progesterone and pregnenolone, as well as the increased metabolism of the utero-placental tissues, demonstrating that the foetus responds to stress (Ousey et al., 2005). Low and stable levels of corticosteroid are optimum for the development of brain areas sensitive to stress and glucocorticoids – too high or too low causes problems for the foal later in life.

Breeders often provide a close analogy to the free-living environment, whether they are a hobby owner breeding just one foal, or whether they breed many horses and keep their mares in mutually compatible groups. These practices protect the mares against significant stress, and the chronic, grinding stress of being housed alone with little or no turnout. Such restrictive and lonely management is akin to the so-called "puppy farm" where lack of stimulation for normal behaviours, as well as lack of daylight, creates chronic stress. The ability to enjoy friends, forage and freedom, (the three Fs coined by equine behaviourist, Lauren Fraser, Fraser, 2012), is paramount to creating a platform for resilience in both the mare and the unborn foal. With this buffer against the negative stresses that life presents, the mare can easily recover her equilibrium and remain mentally healthy.

Foaling: the new arrival

This is the most eagerly anticipated moment! I once worked as a stud groom where the mares were brought to a nursery paddock directly outside my employer's house. He was able to look out of the window first thing and check for new arrivals, and that essentials like expulsion of the placenta had occurred – something to be quick to check for as the fox can be partial to a bit of afterbirth! Other breeders take a more scientific approach; Jackie Hughes fitted her mare with a foaling alarm from two weeks prior to the estimated due date, so that she could be alerted to the pending birth.

Labour in horses is relatively quick. When the mare feels the first contractions, she typically makes herself safe, situating herself close to her group, but not always in the middle of it. It's important to horses not to attract the attention of predators, so many mares foal down in the dead of night, however this is not a hard and fast rule. A hill pony mare who is comfortable in her environment will just as easily foal in the middle of the day. After avoiding predation, the next important thing is for the mare to be the one closest to the foal as it stand ups and looks for a dark place between two uprights to suckle. Foals can be seen to try between the mare's front legs, and even the corner of the stable. Once the mare has carefully sniffed and licked her foal, identifying him as her own, she will then progress to gently guiding her foal to her udder. If she perceives too much interference from outside – other horses or people – then she might behave aggressively to drive them away.

Clearly, choosing the foaling environment is important to facilitate success. Foaling a mare down in a group of horses is possible but runs the risk of other brood mares near their due date trying to steal the foal if the mix of mares and predicted due dates are not carefully managed. A high-ranking mare will not allow her baby to be "stolen", but a lower-ranking mare who foals first can be in for a stressful time as she tries to keep her foal close.

In free-living herds, it is fairly typical that dominant mares conceive before subordinate mares, and consequently foal before them too. Naturally, this is a chicken-and-egg conundrum; this situation might come about because dominant mares have less stress in their lives, and are in better physical condition than other mares, so might be more attractive to the stallion as well as being able to present themselves to him ahead of their subordinates (Curry *et al.*, 2007).

Mary Prewhitt takes great care in how she organises her mare groups when foaling down in the field. Mary prefers to have a more dominant mare foaling first, as if a less dominant mare foals earliest, sometimes the more dominant mare will try to steal the foal. Mary explains:

> I do like to have a second mare in the pasture if possible during foaling, as that second mare often takes on a midwife role and protects the foaling mare. In my location, we do have large predators, so having a midwife mare is a safety consideration in a large pasture. It's also helpful for the foaling mare, if she tends to be anxious she will allow a midwife mare to attend her. I have one mare in particular who is a wonderful midwife mare, even for mares who are less dominant than she is. She's worth her weight in gold.

Mary also allows her mares to foal in the pasture with the stallions, weather permitting. She believes that, if not confined in a stable, the mares can walk and pace during labour, enabling a more comfortable delivery. She describes the stallion's behaviour during foaling:

> I've never had a stallion be pesky or try to be involved in this process. Generally they stay on the other side of the pasture and leave the mare to "woman's business". Generally within a few hours of foaling, the stallion will come down and introduce himself to the new foal.

Mary's stallions know their mares extremely well and have perfect understanding of fatherhood.

Where there is any doubt about the rest of the herd's ability to properly cope with the new arrival, mares can be foaled down in paddocks or other areas immediately adjacent to their usual social groups, provided they are made familiar with them and can relax in those areas beforehand. I've experienced mares foaling in a large area of their home field sectioned off with temporary fencing; the rest of the group come to meet the new foal, and when all are settled, usually in a week or more, the fencing has been removed and the group reintegrated as normal.

Growing a baby horse

Once the foal has located the udder and fed for the first time, its ability to identify its mother and stay close to her gradually increases. The mare will safeguard this process by staying close to her foal at all times for the first two weeks or so, to prevent the foal from wandering off and making mistakes (Houpt, 2002). In practice, foals younger than a few weeks will not necessarily follow their mother when she is led away, and if the two become too far apart the mare will become distressed and attempt to get back to her baby. This can be dangerous to her human handler. If mares and very young, un-halter-trained foals have to be moved, then the distance to be covered should be as short as possible while also allowing the two to remain in line of sight, and a lone handler might well have to turn the mare back to the foal in order to "collect" it from its wanderings.

When a group of horses contains only a smaller proportion of mares with young foals at foot, it can be typical for the mares with foals to keep themselves together in a small nursery group (van Dierendonck et al., 2004). It is likely that this aids maintenance of the mare–foal bond as well as enabling young foals to develop their social skills with their peers. Prior to one month of age, foals generally only direct play towards their self or their mother; after one month of age they begin to interact more widely (Crowell-Davis et al., 1987). Play between fillies, or fillies and colts, is normally concentrated around grooming and chase games, whereas between colts there is far more play fighting. In Figure 2.3, a younger colt wrestles with a slightly older one, who is self-handicapping to encourage his younger brother in the play fight.

It logically follows that mares with young foals might either need to be kept separate from any main herd, be kept in a band of other mares with similar-aged foals or be kept in an area large enough for them to seek and maintain distance from other horses should they so wish. As soon as the foal is interested in more social play, there need to be other socially skilled horses available for him to play with. To keep a mare and foal on their own can make life hard for the mare; some foals, typically colts, are very playful and it's tiring for mares who not only have to produce milk for them, but also keep watch over them without the help of other adults, and on top of that, have to keep up with their robust play which, in the absence of more suitable playmates, they might also direct to people, sometimes sadly leading to being punished by the human and creating fear associations that exaggerate the issue.

Figure 2.3: Colts enjoy participating in rough and tumble play. Photo: Jenni Nellist.

> **Case study: JJ and Usha**
>
> Jackie Hughes kept her mare and foal separate for the first three months, just over the fence from the usual group, and then let them in with the others – all geldings who undertook their uncle roles in playing with the colt.
>
> > JJ was an independent foal and Usha wasn't an overprotective mum. I put the horses out in their separate turnouts until lunchtime, then opened up a fresh field that had been rested for the three months since JJ was born. I allowed the mare and foal out first, then allowed the geldings to join them. There was lots of fresh grass and Usha was happy to get her head down. JJ introduced himself to the geldings in quite a quiet way showing mouthing behaviour, the geldings gave him a good sniff and a little canter around and then they all settled to eat the fresh grass. Ash was very bonded to Usha and she was happy to mutual groom with him. Chester was happy to take over foal duties grazing next to him and sharing a tail with him. After a little while, Ash moved over to JJ and shared a tail with him allowing him to graze close to him. Usha was quite happy to hand over the foal for the geldings to look after and was not overprotective of him. JJ was a confident and fairly independent foal, none of the herd showed any aggression, neither did the foal; they were rotated onto fresh grass every three to four weeks ensuring there was no shortage of food for the herd. They quickly settled back into a herd sharing babysitting duties .They were all stabled at night.
>
> Jackie paid great attention to her colt's education, handling him gently from birth for short periods of time and taking care that he could enjoy the attention without it becoming overwhelming or overexciting. She recalls:
>
> > *The foal was handled gently from birth, just getting used to having a person around only for very short periods. I was really proud that when visitors came he cantered over to see them. He will still come to call.*

Early training

The other important part of growing a foal is to train them to be haltered and to cooperate with handling by people, such as for veterinary procedures and hoof care. Some foals are inherently more fearful than others and find these experiences intrusive and frightening. They need more time, patience and carefully structured training to help them cope with human expectations. Good halter training prevents injuries and fear that can persist for a lifetime, so looking for and carrying out the best practice is important.

First and foremost, foals are generally curious animals. If their mother is comfortable with people around them, then foals will normally come and investigate their human visitors. Most then very quickly learn to seek out enjoyable human scratching. Scratching the withers of a foal will normally have the result of lowering their heart rate and boosting their bond towards humans (Feh and de Mazières, 1993). Since this is such a nice activity, the next lesson is normally in moderating what behaviour from the foal will earn the scratches, hand in hand with signals for when there are no more scratches to be had. Desiring human attention,

> ### Case study: childcare
>
> At Mary Prewhitt's, the stallions live with their families during the summer and early autumn. In Mary's experience, the stallions act as a father role for the foals whether or not they are genetically theirs and she has often seen stallions take on childcare duties.
>
> > I've seen stallions take the older foals off down the pasture, leaving the mares up top socialising. Stallions often seem to enjoy playing with the foals, more than mares do. They play bitey face and tag and seem to tolerate much more "naughty" behaviour from the foals than mares do. I wish I could say that I see a difference in foals that were raised with stallions present as opposed to an all-female herd, but I honestly can't. The ones that I see who really benefit from this set-up are the stallions – who get to live natural and happy lives.

but not getting it immediately can cause some foals to become frustrated, and in their frustration they might behave aggressively or in an excessively playful way. Entering the field or other enclosure when the foal(s) are settled, and then approaching them once the mares have received some attention and the fences have been checked, helps them to learn that calm foals get a desirable human approach. It's then instructive to them to commence scratches when they are not actively trying to coerce you into this activity, such as by reversing up to you! Finishing with a clear "end" signal and wandering around to check the field once more before leaving usually helps foals to learn the benefits of keeping calm.

Once the foal has learned the benefits of a good scratch, this reward can be integrated into necessary husbandry training; to wear a halter and be led, to be groomed, to cooperate with hoof care, and to cooperate with the vet for vaccines and (in the UK and Europe) the compulsory microchip.

Orphans

Even with the best-laid plans, a mare can die during birthing or afterwards, or reject her foal, and you are left with a foal that you need to help – literally or functionally an orphan. In these cases, the most important thing is that the foal grows up a horse and not as a hairy human. A person with a busy schedule, with time only available for feeding and little for socialising, in combination with a small herd of socially adept horses often prove to be the best medicine.

If a foal is orphaned in the "wild", under feral conditions, then provided they are old enough to survive without milk and can latch onto and keep up with a family group, then they will grow up successfully as part of the herd. A few years ago, I observed one such colt, aged approximately 3 months, who couldn't be caught when his mother was found ill and dying. He hung around the site of his mother's capture for a few days and became separated from the rest of his mother's group. Eventually, he began to wander and was later found trailing along with a group of four mares and their foals of a similar age. The mares were not very tolerant of the young interloper, but the colt was able to live peacefully on the fringes of the group and was often seen grooming and playing with the other foals.

When a domestically raised foal is orphaned there are two options: a foster mare or hand rearing. It is possible to foster an orphaned foal onto a mare who has lost her own foal provided this is done quickly, before the mare's maternal instincts wane and her milk dries up. Best practice is to swap the scent of the mare's foal onto the orphan as mares recognise their foals by scent, and this increases the likelihood that the foster mare will accept the orphan. I have seen semi-feral mares who have lost their foal, co-raise the foal of another in their group, leading to the lucky foal having two milk bars and the mares being able to keep better condition too.

Where there is no option for a foster mare, good practice is to feed formula milk (buy large quantities so you do not need to change the type), deliver it at body temperature and always from the same teat – foals like to keep their teat! Regular feeds at two-hourly, then three-hourly, then four-hourly intervals combined with the gradual introduction of creep feed as the foal ages, makes sure they adapt and learn to cope with the feelings of frustration around disruption of their natural feeding ethogram. Keeping such orphans with peaceful adult horses and those who will play with them helps them to feel safe and to learn proper social skills. Consistent training to be haltered, led and to cooperate with hoof care is also important and prevents orphans from seeing humans as playmates and directing unwanted play behaviour towards them – such behaviour is often sadly misinterpreted by humans, which can lead to humans punishing the foals, who become frightened, frustrated, anxious and develop insecure attachment bonds to people.

Weaning

Perhaps unexpectedly, it is the youngster who cuts the apron strings, and sometimes this doesn't happen at all. Normal attachments between mother and young occur in order to provide the young with a sense of security, which supports their exploration of the world around them. The support comes from being able to return to mother at any point when emotional distress is felt.

Naturally, weaning is a significant life event, and in horses, it comes in two parts; nutritional weaning where the foal no longer needs to nurse, then the foal leaving the mother and the herd completely and striking out in a new group. Foals feed from their mothers for nutrition and to develop attachment – to return to a feeling of comfort and safety via the appeasing pheromones produced by the mare. Feral mares normally stop suckling their foals when the foals are around 11 months of age. Mares will tend to reject their foal's advances around a month prior to giving birth to the next foal; lactation and the last stages of pregnancy come with a high energy cost and so nutritional weaning will normally take place at this stage. If a feral mare is not expecting another foal, she might continue to tolerate her now juvenile offspring nursing, and this will often happen because of some emotional disturbance rather than just to feed. I've seen a yearling hill pony filly return to nurse when there had been an argument between two rival groups over bread dished out by well-meaning passers-by; she clearly needed to increase her sense of security (see Figure 2.4).

The juvenile horse will finally become completely independent of mother somewhere between 18 months and 4 years as it voluntarily leaves the herd in search of its own mates. Contrary to popular belief, young horses are rarely expelled from the herd; colts normally leave with their peers to look for mates, and fillies may experience increased aggressive behaviour from their fathers if they have not hooked up with other stallions. Where there is an unnatural ratio of mares to stallions such as

Figure 2.4: Cefn Hugo Boss Femme's yearling filly returns to nurse after a disturbance between neighbouring groups. Photo: Jenni Nellist.

in UK semi-feral herds, daughters will remain with the mothers even for life. I have witnessed a mare and stallion living with their five sons, aged from a foal to 4 years of age, the older three colts frequently roaming away from the group, then returning again.

So, how does this compare with foaling in domestic herds? Typically, in stud farms foals are artificially weaned – an enforced separation of mare and foal at an earlier age and/or in more forced circumstances than would occur in the wild. Such abrupt separation predictably has some negative effects, which need to be avoided as far as possible. Early and abrupt weaning not only causes a nutritional stress, but a significant psychological one too – the foal has lost his secure base and becomes distressed. Abrupt weaning is associated with the onset of crib-biting in horses with the genetic predisposition to do so, and also with gastric ulcers. Early, abruptly weaned horses also tend to become depressed post weaning, or alternatively, hyper-vigilant and lacking in sleep as without mum, they feel vulnerable and defensive. These negative effects of bad stress can have effects permeating throughout life as the horse is now either more sensitive to stress, or the stress response is dulled, leaving the animal depressed.

It is not also only the foal that can become stressed by acute and early separation. The mare knows that it is her job to keep her baby safe from harm. When feral foals wander off, away and out of sight, the mares will call until they are reunited, often with the help of the stallion who will actively go in pursuit of the foal and herd him or her back to the group. When a mare can't locate her foal, and knows he/she has not died, she will become distressed and hyper-vigilant, more so if she is completely isolated. Abrupt, early weaning therefore causes the mare mental anguish and frustration of behavioural goals associated with bringing up her young.

How could horses be weaned, with behaviour in mind, in the domestic business setting? When and how to wean are two decisions that affect the quality of life of mare and foal. Puberty usually onsets in the spring after birth, but sex hormones can begin increasing earlier – as early as 6 months of age in Thoroughbreds. Puberty involves increases in emotional behaviour and a higher risk of forming fear associations – coincidence of puberty and artificial weaning procedures need to be carefully managed. A behaviourally-minded approach would include delayed artificial weaning from the traditional 4 to 6 months of age to the more natural and normal 10 to 11 months of age, and to also follow a lower stress protocol. We know that foals weaned in groups onto pasture and not given concentrate feed, results in lower incidence of stereotypical behaviours (Erber *et al.*, 2012). The presence of familiar adult horses will also help to maintain the foal's sense of safety. It might be tempting, with a large group of mares and foals, to remove one or two mothers each day. However, from both the mares' and foals' point of view, this could be a prolonged and unpredictable situation ("Who will be next?"), and thus likely to increase stress – this was found to indeed be the case in a study by Erber and colleagues.

Signs to look for when considering weaning include:

- the foal's increased independence from the mother;
- foal spending more time with other herd members;
- the mare losing her inclination to call for her foal when out of sight;
- less frequent and shorter nursing bouts;
- the mare sometimes preventing nursing bouts.

Looking after adolescents

Puberty can be just as challenging to "teenage" horses as it can be to human teenagers. Across all the mammals, emotionality and the processing of rewards peaks at puberty, while the cognitive processes will not be mature until adulthood (Brenhouse and Andersen, 2011). In practical terms, young horses are more excitable, can't think very clearly and lack inhibition. This renders them more prone to feelings of fear and frustration, especially in response to threatening behaviour from other horses and from people. For example, when confused by their handler, some young horses try to appease them with play. Unfortunately, their choice of play can be too rough, and this presents as dangerous, possibly aggressive behaviour to the handler. In turn, this can tempt some people to resort to the use of physical punishment, which might serve to inhibit such behaviour, but which might alternatively provoke aggressive behaviour in self-defence depending on whether the youngster's innate

Case study: JJ's weaning

Jackie Hughes was able to wean her colt in a gradual manner. Since he was bred to be her next riding horse, he continued to be kept at home with the rest of the group including his mother without ever being a problem to take out, for example for riding.

> JJ weaned naturally; stabled at night in his own stable from 1 year old, mum gradually withdrew the milk bar herself during the day when he was 18 months old. I left it to happen naturally as in the wild.

tendency is to be proactive in response to stress and fear, or to withdraw and become depressed.

Not only are young horses more susceptible to fear but they are also susceptible to experiencing frustration when anticipated rewards fail to materialise, or when kept in an under-stimulating environment. Sudden isolation at 2 years of age by use of single stables has been demonstrated to induce depression and blunting of the stress response (Visser *et al.*, 2008). Clearly, great care is required when managing, handling and training the young horse.

Puberty starts in the spring after birth when day length increases; this means that some late foals may reach puberty at an earlier age than foals born before them. Sex hormones have been seen to increase in Thoroughbred foals as young as 6 months of age, then again during the months of spring through to mid/late-summer when they become yearlings (Dhakal *et al.*, 2010). In other populations, puberty may start when the horse reaches 2 years of age; around 50 per cent of Brazilian sport horse fillies and 50 per cent of Lipizzaner fillies had their first ovulation at around 25 months of age (Souza *et al.*, 1997; Čebulj-Kadunc *et al.*, 2006). Puberty spans an age range from 6 months at the very earliest to onset during the 2-year-old year at the latest, with oestrus lasting longer in 2-year-old fillies than in yearlings, indicating the greater advance towards social maturity which is nearer to 4 to 6 years of age.

In the natural world, colts and fillies in free-living populations live in the group in which they were born, until around the age of 2 years or older. No doubt living in a safe and stable social environment helps to support them through puberty. Although some are still adolescents when they leave the stability of the family group, it is still largely their choice as they make explorative forays into the environment surrounding their parents' home range. Emotionality has been found to be consistent within individual horses from the age of 5 months up to 2 years, and those that received consistent training over that period were concluded to be less emotional (Visser *et al.*, 2002).

In the domestic setting, keep young horses in mutually compatible groups in order to protect against the negative effects of stress and to promote social skills and cognitive development. It is important to handle and train young horses frequently and consistently for short periods of time per day, accommodating their social and cognitive development and applying learning theory correctly.

Buck: two years later

Buck was introduced to covering mares in hand, and at first took a long time to court the mares and breed them, but very quickly became very impatient and aggressive to the mares and the handlers. His dangerous behaviour was making it hard to get mares covered safely, and caused them and the handlers much frustration and distress. Since Buck's behaviour was a health and safety issue as well as an injury risk to both himself and the mares, Dave decided to try only breeding Buck in the pasture and for the next breeding season, Buck was turned out with a small number of experienced brood mares before they came into oestrus. Since then, Buck has relearned how to court mares in a more natural manner and serves both at pasture and in hand, taking his time to court mares and is far quieter to handle, making life safer for Dave and his grooms, and more productive for him and his mares as covering is no longer a stressful or poorly timed activity now that a behaviourally-minded approach has been taken.

Final thoughts

The nature and nurture of breeding horses is to accommodate their natural behaviour and limit the extent to which those behaviours are frustrated, creating a nurturing environment that does not place demands in excess of what horses can successfully cope with. By taking a behaviourally-minded approach, keeping mares and stallions in social, naturalistic living conditions, arranging conception so that normal courtship behaviour can be performed, and carefully raising foals through weaning, puberty and on to adulthood, the breeder not only produces a desirable horse but also promotes high welfare standards. In the modern age, where horse welfare concerns ever more widely influence the public perception of our use of horses, shining examples of practical approaches that protect and support the mental lives of horses are in high demand, to illustrate that by keeping behaviour in mind is the future of sustainable horse keeping.

Further reading

Davies Morel, M.G.C. (2015) *Equine Reproductive Physiology, Breeding and Stud Management* (4th revised edition) CABI, Wallingford, Oxfordshire, UK.

References

Allen, W.R. and Wilsher, S. (2018) Half a century of equine reproduction research and application: a veterinary *tour de force*. *Equine Veterinary Journal*, 50: 10–21.

Boyd, L. Scorolli, A., Nowzari, H. and Bouskia, A. (2016) Social Organization of Wild Equids. In Ransom, J.I. and Kaczensky, P. (Eds), *Wild Equids: Ecology, Management, and Conservation*. Johns Hopkins University Press, Baltimore, MD, USA. p. 7.

Brenhouse, H.C. and Andersen, S.L. (2011) Developmental trajectories during adolescence in males and females: a cross-species understanding of underlying brain changes. *Neuroscience Biobehavioural Review*, 35(8): 1687–1703.

Cebuli-Kadunc, N., Cestnik, V. and Kosec, M. (2006) Onset of puberty and duration of seasonal cyclicity in Lipizzan fillies. *Equine Veterinary Journal*, 38: 350–353. DOI: 10.2746/042516406777749137.

Crowell-Davis, S.L., Houpt, K.A. and Kane, L. (1987) Play development in Welsh pony (Equus caballus) foals. *Applied Animal Behaviour Science*, 18(2): 119–131.

Curry, M.R., Eady, P.E. and Mills, D.S. (2007) Reflections on mare behaviour: social and sexual perspectives. *Journal of Veterinary Behaviour*, 2(5): 149–157.

Dahkal, P., Otsuka, A., Nambo, Y., Harada, T., Nobuo, T., Itoh, M., Gen, W. and Taya, K. (2010) Dynamic change in circulating pituitary and gonadal hormones from birth to puberty in the Thoroughbred colts and fillies. *Biology of Reproduction*, 83(Suppl. 1): 580.

Dindot, S.V. and Cohen, N.D. (2013) Epigenetic regulation of gene expression: emerging applications for horses. *Journal of Equine Veterinary Science*, 33(5): 288–294.

Erber, R., Wulf, M., Rose-Meierhofer, S., Becker-Birk, M., Mostl, E. and Aurich, J. (2012) Behavioural and physiological responses of young horses to difference weaning protocols: a pilot study. *Stress*, 15(2): 184–194.

Feh, C. and de Mazières, J. (1993) Grooming at a preferred site reduces heart rate in horses. *Animal Behaviour*, 46(6): 1191–1194.

Fraser, L. (2012) The horse's manifesto: what do we want? Friends, forage and freedom! https://iaabc.org/horse/the-horses-manifesto-what-do-we-want-friends-forage-and-freedom-part-1-of-3 (accessed 26th February 2018).

Freymond, S.B., Briefer, E.F., von Niederhausem, R. and Bachmann, I. (2013) Pattern of social interactions after group integration: a possibility to keep stallions in group. *PLosOne:* https://doi.org/10.1371/journal.pone.0054688

Houpt, K.A. (2002) Formation and dissolution of the mare–foal bond. *Applied Animal Behaviour Science*, 78(2–4): 319–328.

Kimura, R. (2001) Volatile substances in feces, urine and urine-marked feces of feral horses. *Canadian Journal of Animal Science*, 81: 411–420.

Lee, J.R., Hong, C.P., Moon, J.W., Jung, Y.D., Kim, D.S., Kim, T.H., Gim, J.A., Bae, J.H., Choi, Y., Eo, J. and Kwon, Y.J. (2014) Genome-wide analysis of DNA methylation patterns in horse. *BMC Genomics*, 15(1): 598.

Malschitzky, E., Pimentel, A.M., Garbade, P., Jobim, M.I.M., Gregory, R.M. and Matos, R.C. (2015)

Management strategies aiming to improve horse welfare reduce embryonic death rates in mares. *Reproduction in Domestic Animals,* 50: 632–636.

McDonnell, S.M. (1992) Normal and abnormal sexual behavior. *Veterinary Clinics of North America: Equine Practice,* 8(1): 71–89.

McDonnell, S.M. (2000) Reproductive behavior of stallions and mares: comparison of free-running and domestic in-hand breeding. *Animal Reproduction Science,* 60–61: 211–219.

McDonnell, S.M. (2005) Sexual behaviour. In Mills, D. and McDonnell, S. (Eds), *The Domestic Horse: The origins, development and management of its behaviour.* Cambridge University Press, Cambridge. pp. 110–125.

McDonnell, S.M. (2008) Practical review of self-mutilation in horses. *Animal Reproduction Science,* 107(3-4): 219–228.

Ousey, J.C., Houghton, E., Grainger, L., Rossdale, P.D. and Fowden, A.L. (2005) Progestagen profiles during the last trimester of gestation in Thoroughbred mares with normal or compromised pregnancies. *Theriogenology,* 63(7): 1844–1856.

Painter, R.C., Roseboom, T.J. and Bleker, O.P. (2005) Prenatal exposure to the Dutch famine and disease in later life: an overview. *Reproductive Toxicology,* 20: 345–352.

Powledge, T. M. (2011) Behavioral epigenetics: how nurture shapes nature. *BioScience,* 61(8): 588–592. https://doi.org/10.1525/bio.2011.61.8.4

Souza, J.A.T., Gacek, F., Oliveira, J.V. and Augusto, C. (1997) Growth and onset of puberty in female Brazilian Sport Horses. *Revista Brasileira de Reprodução Animal,* 21(3): 117–120.

Van Dierendonck, M.C., Sigurjonsdottir, H., Colenbrander, B. and Thorhallsdottir, G.C. (2004) Differences in social behaviour between late pregnant, postpartum, and barren mares in a herd of Icelandic horses. *Applied Animal Behaviour Science,* 89(3-4): 283–297.

Van Niekerk, C.H. and Morgenthal, J.C. (1982) Fetal loss and the effect of stress on plasma progestagen levels in pregnant Thoroughbred mares. *Journal of Reproduction and Fertility,* Supplement 32: 453–457.

Visser, E.K., van Reene, C.G., van der Werf, J.T.N., Schilder, M.B.H., Knaap, J.H., Barneveld, A. and Blokhuis, H.J. (2002) Heart rate and heart rate variability during a novel object test and a handling test in young horses. *Physiology and Behaviour,* 76: 289–296.

Visser, E.K., Ellis, A.D. and van Reenen, C.G. (2008) The effect of two different housing conditions on the welfare of young horses stabled for the first time. *Applied Animal Behaviour Science* 114: 521–533.

Wilsher, S.A. (2009) Studies in equine reproduction. University of Bedfordshire. Accessed online 06.01.2018: http://uobrep.openrepository.com/uobrep/bitstream/10547/134931/1/wilsher.pdf

Yarnell, K., Hall, C., Royle, C. and Walker, S.L. (2015) Domesticated horses differ in their behavioural and physiological responses to isolated and group housing. *Physiology & Behavior, 143*: 51–57.

Chapter 3
Training
Catherine Bell

This chapter considers the training of horses in a way that enhances their welfare. Such training requires an understanding of how horses learn, what motivates them and how to recognise and avoid equine stress. Instead of simply telling the reader how to train a horse, the emphasis is on the horse's experience of being trained and how this can be improved.

The past 30 years have seen significant change in the styles and philosophies of horse training. In the UK, the divergence from traditional styles of training arguably started in 1989 when, invited by the Queen Mother, horse trainer Monty Roberts' visit was televised. Equitation in Europe had been previously rooted firmly in military training – to this day, we mount from the left side so our swords do not impede our ascent – but, helped along by the Internet, a variety of alternative styles has now reached our shores. Different versions of "Natural Horsemanship", ostensibly a "new" approach but strongly influenced by old Western cowboy methods, opened our eyes to alternatives. Following in its wake, we also had the more science-based clicker training, a method born in the science laboratories of the 1950s and popularised via the training of captive marine mammals. The debates for and against the various methods have been extensive and ongoing and it would be easy to spend this chapter debating the advantages and disadvantages of each individual method. However, it is perhaps more useful to look beyond the method and, keeping behaviour in mind, consider the experiences of the horses undergoing the training. Thus, we equip ourselves with the means to assess the merits of any method that we may come across in the future and make informed choices about the activities in which we engage with our horses.

 A horse called Jasper

Jasper was a horse who had not had his feet trimmed in a couple of years. Seemingly caused by a painful mite infestation in his feathers, he had become increasingly aggressive to handle and would no longer let anyone, including his owner, touch his legs. He had reared and kicked each time the farrier had attempted to trim his feet, ultimately with the farrier (understandably) declining to try again. A vet had received a similar response, with Jasper rearing and "boxing" at the vet with his front legs. Whereas some horses could cope with allowing their feet to "self-trim" naturally and not need the intervention, Jasper was not one of these. He had multiple cracks to his hoof wall and significant flare and painful infection that needed addressing. He needed to have his feet trimmed as quickly as possible but in a manner that did not cause him additional distress and trigger his dangerous – albeit defensive – behaviours.

I just want to ride my horse. Why do I need to know about behaviour?

Behaviour is everywhere. Traditionally, if you were a "good rider" it meant that whatever the horse was doing or feeling, you could stay on and make the horse perform. But what if knowing about behaviour could enhance your riding and your horse's performance, not to mention keep you safer? Have you ever tapped a horse with a whip on approaching a jump "just to make sure"? Have you ever given a horse some food "just to shut him up"? Have you ever wished your horse had better "manners"? Have you ever thought a horse was "testing your boundaries" or "taking the Mickey"? If so, then it is likely you would benefit from an exploration of horse behaviour.

When faced with a horse like Jasper, it is easy to think in terms of "what do we need to do in order to stop the aggressive behaviours?". Or to take the attitude "I know you're in pain, but I have to do this". As we shall see, this is how some training methods do indeed address problem behaviours and, if applied effectively, they can appear to work. At least, from the human's perspective, the problem behaviours can cease and leave the human no longer troubled by the horse's behaviour. But what is the horse's experience when we focus purely on getting rid of behaviours that are inconvenient for the human?

Any time a horse is doing something we don't want, or not doing something we do want, then the first thing we should do is to stop and think about what is happening. What is different between the occasions when the horse is behaving in the undesirable manner and the occasions when the horse's behaviour is non-problematic? It is a difficult truth for us to accept but commonly the presence of a human – or the human behaving in a particular manner – is that crucial factor. After all, the horse is not, for example, biting when we are not there to be bitten. What has changed? And why?

The second thing to do is consider the cause of the behaviour, or lack thereof. Sometimes we are lucky and there is a clear, sole cause that can be addressed and the problematic behaviour instantly corrects itself. More commonly, the cause is multifaceted and we need to look at everything in turn. The different elements we must consider include genetics, history, management, pain, confusion and fear.

Genetics certainly plays a part in causing problems but, since there are no individual genes governing specific behaviours, we cannot rely on genetics in isolation to explain away an unwanted behaviour or response to one of our requests. The genetic make-up of the horse will play a large part in determining how readily he responds to training, for example his resilience to training regimes and his fear of stimuli in the environment. Thus, genetics is a relevant factor but not one we can change; instead, we need to consider the horse as an individual and ensure that we are meeting his emotional, physical and behavioural needs at any given time. Likewise, the horse's history will have an impact on his behaviour in the present and future. Although we cannot change a horse's history, even horses who have suffered extreme neglect and abuse can still be rehabilitated to some extent; again, the crucial factor is to meet the needs of the individual horse and not attempt to mould the horse to any human-imposed agenda. The horse's environment and management play a large part in contributing to a horse's behaviour, as we saw in Chapter 1, and if problems relating to management are not addressed, then the horse's experience of, and attitude towards, training can be significantly impacted.

Once management issues have been addressed, when we consider the cause from the horse's perspective we find that most behavioural issues are underpinned by pain, fear or confusion. Pain issues require the services of a veterinarian and/or other equine health care practitioner and thus are beyond the scope of this chapter. Once the horse is free of pain, he can be considered ready for training.

Fear and confusion, however, are directly relevant to how our training should proceed. A novice horse who has not yet begun his "education" cannot be expected to know innately the standard cues and aids that we tend to use in training. It is common for young horses to be left at grass to "be a horse" for their early years and then to be brought into work when they are old enough to be backed for riding. Simple skills such as preparation for leading, hoof handling, touch acceptance and exposure to novel objects are commonly overlooked in favour of ridden work, with the result that the horse can find even these basic tasks frightening and confusing. Aids and experiences need to be taught or the horse will feel distress from being the recipient of instructions that he does not understand. If the horse's genetics and/or history mean that the horse is finding these experiences frightening as well as confusing – and if we consider that the horse is a prey animal and so is predisposed to find the unknown frightening – then we are setting him up to fail before we even start.

Both fear and confusion can be largely avoided if we make use of a process called shaping. This allows the systematic introduction of new behaviours and experiences such that the horse's thresholds for fear and confusion are not triggered (see Figure 3.1).

Shaping

Shaping is the means by which we use progressive approximations of the behaviour to improve gradually, from the horse's current

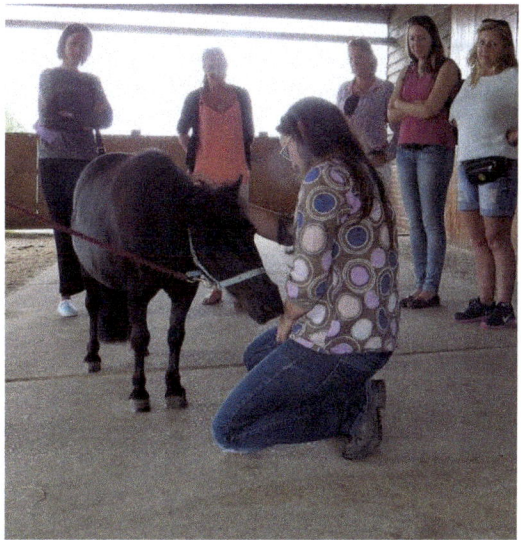

Figure 3.1: Even greeting a horse in a non-threatening manner can be a big "ask". Some small ponies can find tall humans "towering over" them distressing – this pony responded well to the handler's efforts not to do so. Photo: Suzanne Rogers.

ability to our ultimate desired ability. Such an approach applies to a youngster learning new fundamental tasks, a rescue pony recovering from trauma or an experienced competition horse being prepared for work at a more advanced level – and all horses and donkeys in between. At a simple, "common-sense", level it means that we "learn to walk before we learn to run". We teach a horse to lead in hand before we try to ride him; we teach a horse to accept touch in the least vulnerable areas of his body, such as shoulders and neck, before we attempt to touch the vulnerable areas such as belly and legs; we back a youngster by teaching him to accept a small degree of our weight before we mount fully.

But this is not truly shaping. Instead, we can extend these concepts into an "art form" when we ensure that we have incorporated every possible intervening step between starting point and goal. Proper shaping includes many steps, ideally written down into a shaping plan. Proper shaping means that we don't progress to the next stage before the horse has become

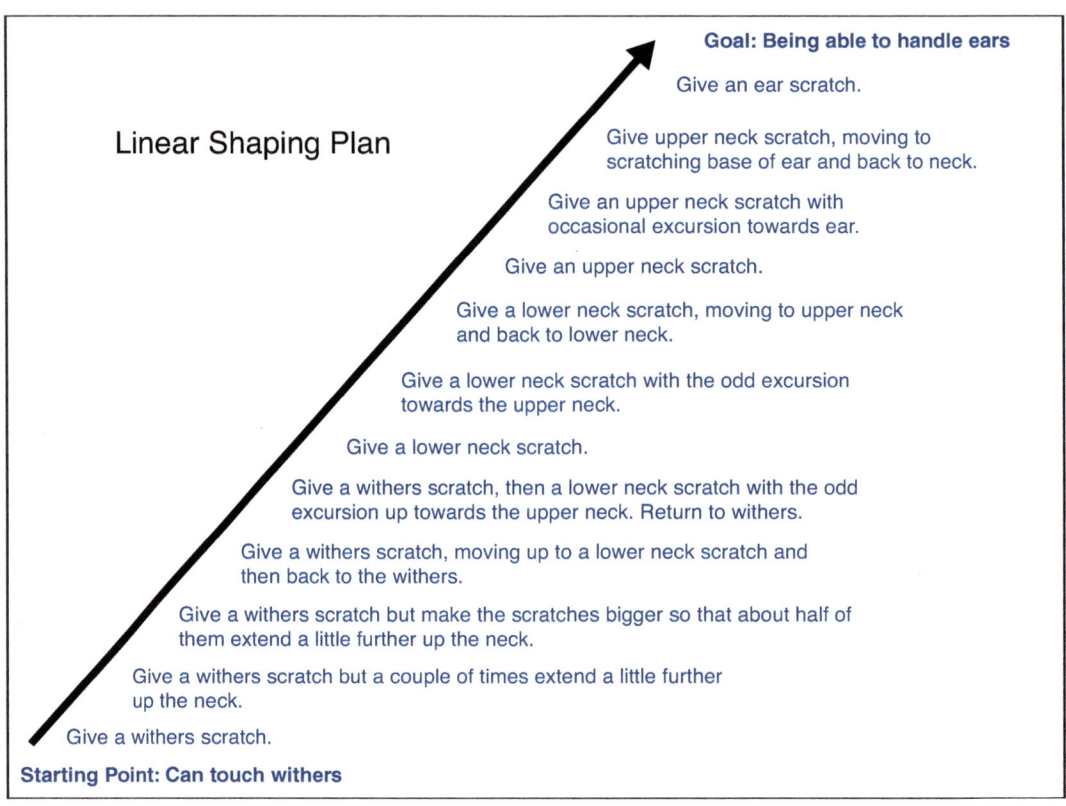

Figure 3.2: An example of a linear shaping plan. This is designed to be illustrative only; individual horses might need variation to suit their needs. Each stage should be conducted on both sides of the horse. After giving an ear scratch, some handling will be possible but further shaping would be needed towards being able to, for example, put cream in the ears.

fully accepting of the earlier stages. We need to be aware of even the tiniest signs of fear and stress, something we will come to later in this chapter, and if the horse displays any of these signs then it is an indication to us that we have moved too quickly. Figure 3.1 gives an example of how even greeting a horse can be included as a shaping step.

Writing shaping plans and keeping track of progress can be an alien approach to training for many horse owners. Behaviour colleague, Maisie Wake, had a novel means of encouraging participation:

If I am facilitating an owner to do this, I am likely to write them out a shaping plan or provide them with a shaping plan template to write out themselves. For private owners, I will sometimes buy them a notebook with a picture of their equine on the front page, named for example, "Missy's Journal". This encourages owners to keep track of their training, and they also have a sense of ownership to the journal because it has been personalised.

An example of how we might use shaping correctly could be touch acceptance (see Figure 3.2). A horse who doesn't like having his ears touched (but for whom a veterinary cause has been ruled out) could be shaped towards allowing his ears to be touched via a straightforward systematic desensitisation programme. Starting a process of stroking at the withers, the strokes could be gradually extended towards the ears. It may take a few strokes, several sessions or even some weeks, before the horse is happy to allow the strokes to reach the ears and so the process should not be rushed. But, by the time of its completion, the horse should be genuinely accepting of the handling, rather than merely being compliant.

A more complicated example might be teaching a horse to load (see Figure 3.3). The shaping plan would be multifaceted, with elements shaping the horse to walk over novel surfaces, walk between close barriers, walk under low ceilings, combine those elements together, load in the home environment, load in novel environments, load in inclement weather, after a hard work-out, and so on. It is so common for people to practise loading at home but then be "surprised" when the horse fails to load after a show. Each element of the shaping plan should be itemised in detailed steps and then combined all together. The shaping plan would need to extend beyond the expected desired behaviour and account for the unexpected extras that come up – this ensures that the horse is prepared for unanticipated eventualities, such as emergency loading to visit the vet.

Karen Pryor, in her book *"Don't Shoot the Dog"* (Pryor, 1984), outlines a series of rules for shaping that provides a useful guide to how to ensure that sufficient steps are included (see Figure 3.4). These rules clarify the manner in which training is rarely linear but proceeds in a series of ups and downs; they explain that the horse may make mistakes, even in elements that he seemed to have understood; and they demonstrate the ways in which humans frequently scupper the horse's learning process. When writing a shaping plan we need to account for these rules and recognise that, for example, following a human over a pole lying on the ground in the arena is a very different

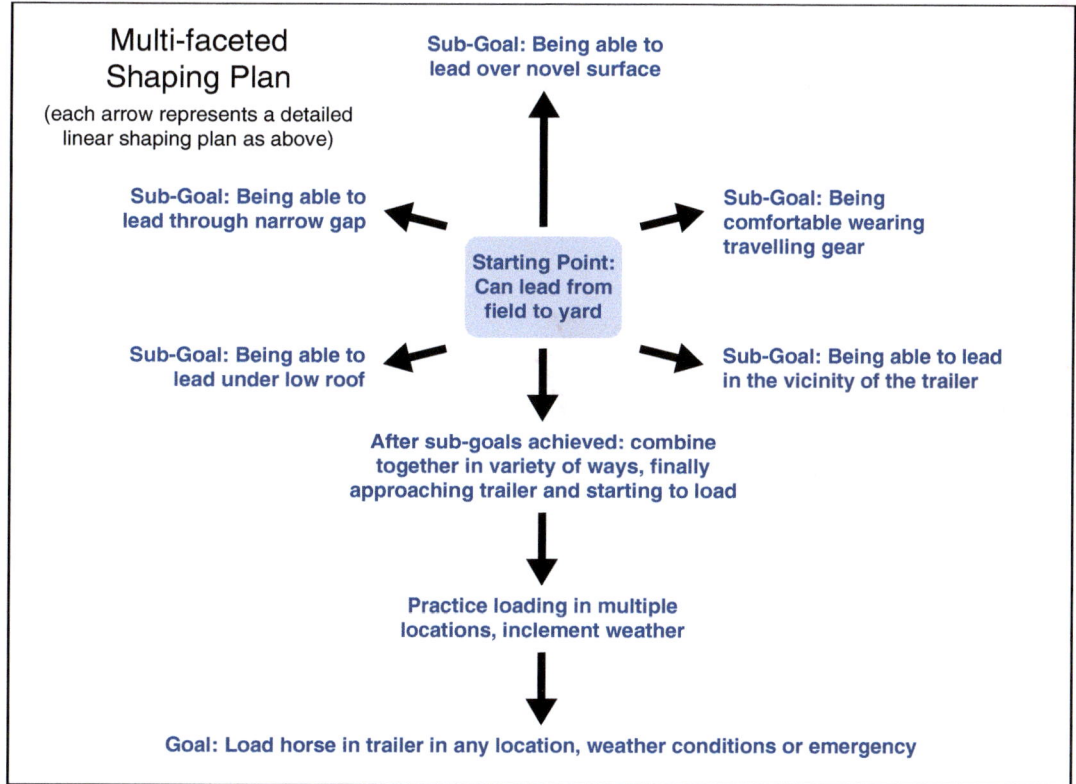

Figure 3.3: Multifaceted shaping plan. Designed to be illustrative only and some horses may need variation to suit their needs.

The Rules of Shaping – Karen Pryor

1. Raise criteria in increments small enough that the subject always has a realistic chance for reinforcement
   - *the horse should barely perceive the difference between two incremental stages*
2. Train one aspect of any particular behaviour at a time; don't try to shape for two criteria simultaneously
   - *plan multiple sessions to address additional criteria (see Figs 3.2 and 3.3)*
3. During shaping, put the current level of response onto a variable schedule of reinforcement before adding or raising the criteria
   - *strengthen a behavioural stage using a VSR before changing the criteria*
4. When introducing a new criterion, or aspect of the behavioural skill, temporarily relax the old ones
   - *previously learnt criteria will likely worsen temporarily as you add a new criterion*
5. Stay ahead of your subject. Plan your shaping program completely so that if the subject makes sudden progress, you are aware of what to reinforce next
   - *planning ahead is key, allowing for the fact the horse may do something unexpected*
6. Don't change trainers in midstream; you can have several trainers per trainee, but stick to one shaper per behaviour
   - *different people train in different ways and switching trainer will confuse the horse*
7. If one shaping procedure is not eliciting progress, find another; there are as many ways to get behaviour as there are trainers to think them up
   - *the shaping plan should be fluid and be adapted to the horse's learning/progress*
8. Don't interrupt a training session gratuitously; that constitutes a punishment
   - *stopping abruptly may be perceived as negative punishment (see later)*
9. If behaviour deteriorates, "go back to kindergarten"; quickly review the whole shaping process with a series of easily earned reinforcers
   - *this can often be a good idea even if all is well, it helps to reinforce the foundations*
10. End each session on a high note, if possible, but in any case quit while you're ahead
    - *keep sessions short*

Figure 3.4: The Rules of Shaping (adapted from Karen Pryor "*Don't Shoot the Dog*").

behaviour from walking over a pole on the ground independently, walking over a pole lying in the field or walking over a tarpaulin. Similarly, it is important not to jump ahead in the programme or to impose rigid or unrealistic time limits. For example, if you have a shaping programme in place to help your horse learn to load then it is not appropriate to have a deadline that requires you to take the horse to a show on a particular date. Such a deadline might allow insufficient time to complete the training; rushing the process could lead to serious consequences, with the horse losing trust in you and regression of progress.

Critics of shaping consider that it is a long-winded process that takes too long due to its inclusion of seemingly pointless steps. But this is the beauty of it. By including many steps, even those that may initially appear unnecessary, you have the opportunity to reinforce the horse for his behaviour, increasing his confidence and understanding. You remove the margin for error, since it may turn out that a step the critic considered unnecessary was actually crucial for the horse. And by removing the margin for error, you reduce the chances of future setbacks, thus saving time in the long-run. Long-time advocate of shaping, behaviourist and contributor of this book, Ben Hart has always believed in the wise investment that is properly planned shaping. He advises "Don't fear small steps, fear not moving forwards" and goes on to say "It's funny how we don't make the time to do it right the first time, but always find time to do it a second time".

Adding to the apparent time-consuming nature of shaping is that in order for it to be most successful it needs to allow for the fact that progress is rarely linear. Instead, it will zig-zag, with improvements some days and set-backs others. I have seen horses who make mistakes, seemingly deliberately, as though to check what happens if they do – that might be an anthropomorphic interpretation but it is hard to see it as anything else when it is a behaviour that is usually well-understood. We also need to progress beyond our desired behaviour so that the horse is not having to perform at his peak level as a norm. For example, if you intend to jump a 1 m show-jumping course at the local show then it would be important for him to feel comfortable jumping 1.3 m in his home environment. Again, rather than making the training too slow, these added elements strengthen the success of the training and make it much more reliable via the added confidence that the horse develops. The progress of future training can often be remarkably rapid once the horse apparently learns to trust that he will not be overfaced.

Shaping Jasper's behaviour

Jasper's feet were in a sufficiently pathological state that there was a genuine need to "get started". But his concerns with hoof care were clearly significant enough for him to resist any rushing on my part and I was anxious not to trigger any of his dangerous behaviours, for his sake or mine. I wasn't sure if I would even be able to handle his feet in the first session, let alone do a proper trim. I still needed to use a shaping procedure.

As I do with most horses, I started with general scratching in non-threatening parts of his body, mainly his shoulders, back, withers and hindquarters. I allowed him to sniff and investigate my hands, clothing and equipment. The scratching was not so motivating for him as to encourage him to seek it but it did allow him to relax in my presence and not worry about what might come next.

As I continued to scratch him, I allowed my hand to make brief excursions down his shoulder and progressively work lower towards his front hooves. The scratches and systematic approach allowed him to find my behaviour predictable and non-threatening. He gave no indication that he found any of this work concerning unless I touched his feathers, in which case he would move and shift his weight anxiously. Consequently, I avoided any contact with his feathers, running my fingers down his leg and skipping from canon bone to hoof before returning to scratches. Repeating this process on both sides, I was gradually able to lift each front hoof but make no attempt to grab hold of them, instead replacing them on the ground. I repeated the process on his hind legs, giving him reasonably pleasurable scratches interspersed with strokes down his legs and avoiding contact with his feathers.

This process was slightly accelerated, given how long it was since he had last had his hooves handled and relative to how long I would have preferred to work with the behaviour. Yet he gave no indication that he couldn't cope with what I was doing and seemed to be enjoying the scratches sufficiently that he was tolerant of my handling. Therefore, because of the state of the hooves, it seemed right to continue but being ready to back off if he gave any indication that he wasn't coping. This is the sort of "real-life" compromise that is sometimes needed, but there is still no excuse to cause the horse any level of distress through excessive rushing.

I returned to the front feet and was able to pick them out, very cautiously and still being careful to avoid contact with his feathers. It would have been easy to think he was "fine" and to accelerate the process at this point. But the reason he appeared "fine" was because I was allowing him to dictate the rate of progress.

Lifting the feet for short periods and allowing him frequent rests was critical. Similarly, I was able to pick out the hind feet, still scratching and still stroking down the legs as I reached for each foot so that Jasper could remain fully understanding of what I was doing. I was then able to repeat each foot, introducing the tools, remove the worst of the excess horn and apply some topical thrush treatment.

It is noticeable that the trim itself was a tiny fraction of the process of helping Jasper become ready for that trim. On future visits, I was able to spend progressively less time preparing and more time on the trim itself. If I had rushed any more – and let us remember that this was already a fairly rushed shaping process on account of his feet needing the attention – then he would not have coped as well as he did. His compliance with the process should not be confused with relaxation or being entirely happy with my requests.

The shaping process did not stop with being able to trim each foot. I also needed to continue the process so that we could still trim in inclement weather and in new locations when he was moved to a new yard. He needed to be able to cope with the inevitable accidental noisy dropping of tools, so we included that in the training. He still needed mite treatment from the vet and it was inevitable that I would sometimes inadvertently touch his feathers; so we needed to practice that a little, allowing for the fact that he was in genuine discomfort as well as probably also reacting from fear of the potential pain.

Gradually, over a year or so, his stress response diminished and I was able to treat him as a "normal" horse, albeit still sensitively.

Avoidance of stress and fear

Accurate shaping relies on the ability of the trainer to anticipate when the horse is about to feel a sense of distress. A horse who feels worried and frightened at any part of his training is likely to demonstrate these emotions through his body language and behaviour. Some of the behavioural indications of stress are well-known to all equestrians – bucking, rearing, bolting, biting and kicking. It is no coincidence that these behaviours are ones that we consider dangerous and to be "behavioural problems". But for the horse, they are simply the last resort defence mechanisms when he finds himself in a stressful situation.

The behaviourally-minded reader will, at this point, recognise that if we could avoid those stressful situations we could thus avoid triggering the production of these dangerous problem behaviours. As a general rule, this is largely true, albeit sometimes easier said than done. However, horses undergoing a mildly stressful experience are unlikely to suddenly engage in one of the big, dangerous behaviours. If they do, then it is debatable whether the horse in question found the experience only mildly stressful. Instead, a mildly stressed horse is likely to engage in a more subtle expression of that stress, for example a spook or step sideways instead of bolting, a fidget instead of a kick, a resistance to ridden aids instead of bucking or rearing, a "grumpy" face such as pinned ears instead of biting (see, for example, McGreevy, 2004). Even these expressions of stress are more explicit than the initial indicators that not everything is OK. The earliest indicators of stress include regions of slight muscular tension, focusing of ears and/or eyes towards the stressor, distraction from the task at hand, triangulated shape of the eye, a "pinched" look to the chin, heavier breathing and/or snorting (see Figure 3.5).

The more we can be vigilant and recognise these initial signs of stress, the less often we will be surprised by a problem behaviour that we incorrectly believe "came from nowhere" (see EBTA, 2012a for video illustrations). It

Figure 3.5a-b: Horses displaying several signs of stress described in the text. (Photos: Natursports and Kseniya Abramova, Dreamstime.com)

is almost always the case that the behaviours do not come from nowhere. A horse will learn through experience that if the trainer fails to notice, or ignores, his early warning signs that all is not OK, there is no point displaying these behaviours. Instead, he will automatically go straight for the more overt behaviours in the first place. In addition to the welfare element, I would argue that the most important thing equestrians can do to preserve their own safety is recognise – and respond sympathetically to – the early warning signs of equine stress. Failure to notice equine stress in horses was identified as being one of the key welfare concerns in the "Horses in Our Hands" report (Horseman *et al.*, 2016).

Returning to our shaping plans, this understanding of the subtle signs of stress, means that if the horse is achieving an early stage of the process but seems "not quite right" or "not really afraid" then he is probably not ready to progress to the next stage. Progression requires that the horse is truly comfortable with each stage, not just achieving the desired behaviour without objection.

Critics might argue that a little bit of stress is not problematic, after all, we all have to deal with a little bit of stress in our lives. Some of us even thrive on it, for example, leaving work too close to the deadline. Although this is true to some extent, there are various factors to consider. First, what humans consider a "little bit of stress" is often very different from the horse's perspective. The horse's behaviour is our guide to how he is feeling – if we dismiss the small signs as "just being silly" then we are merely demonstrating confirmation bias (i.e. we are using the observations erroneously to confirm what we already believe) when we claim that the horse is not really scared. Second, most domestic horses are already enduring a fair amount of stress as a result of inappropriate management (Chapter 1), pain and confusion due to inconsistent training; it is certainly rare that they choose the sources of stress with which they are encumbered. So, a small amount of "extra stress" has a pronounced effect via a process called "trigger stacking", of which more will be discussed in the following paragraphs. Third, by giving the horse the benefit of the doubt we make the process safer and typically progress more quickly in the long term. Finally, if we take a look at the stress processes taking place within the brain, then it helps us to understand that a horse cannot perform to his full potential if he is feeling stressed.

So, what is taking place in the brain at this point? A more detailed description of the physiology of stress is covered in Chapter 8, and also see Sapolsky, 1994 for more in-depth information. However, the subject is introduced here because it is so incredibly relevant to training horses. When a frightening stimulus (i.e. something scary happening) is perceived, the information is received by the brain in the region known as the amygdala, important for the processing of emotions. It releases a hormone called corticotropin-releasing factor (CRF), which triggers the sympathetic nervous system into sending the body into "stand-by mode", such as elevating the heart rate in anticipation of flight, with its greater needs for oxygen and nutrients in the muscles. This can be a quicker response than the more conscious processes taking place via the cortex, which evaluates the true need for a stress response and whether it was a false alarm after all; hence the horse who spooks at "nothing" with the species-appropriate attitude of "better safe than sorry". The CRF also triggers activity in the hypothalamus, which, in turn, releases more CRF that stimulates the pituitary gland into releasing adrenocorticotropic hormone (ACTH). The release of ACTH triggers the adrenal glands into releasing corticosteroids, including cortisol, into the bloodstream.

The release of cortisol is returned via a feedback loop to the prefrontal cortex, hippocampus, hypothalamus and pituitary gland. In a low-stress situation, the cortisol triggers inhibitory responses to reduce the release of all these chemicals back to baseline levels. However, if the stressor is ongoing or multi-element (i.e. trigger stacking) then the continued release of CRF, and its subsequent triggers of ACTH and cortisol, maintain the high levels of cortisol in the blood. In research undertaken on humans and rats, high levels of the chemicals included in this chain of events are associated with depression, anxiety, insomnia, poor appetite and reduced interest in sex (Sapolsky, 1994). They can lead to difficulties in concentrating, learning and memory through their effects on the prefrontal cortex and hippocampus. With their relatively similar mammalian brains, it is logical to suggest that horses experience similar reactions.

In summary, a horse who is stressed – and unable to recover from that stress before the next stressor comes along – is going to struggle to respond to our training requests effectively. When we consider the environmental problems faced by many domestic horses (Chapter 1), we should consider that many horses are very rarely in an "unstressed" state and the addition of stressors via trigger stacking is making life very hard for them. And this is before we consider the level of aversive stimuli included in our training. We come to this next.

How horses learn

The way that animals, including horses, learn is fairly well understood, thanks to research into behavioural psychology that has taken place since the 1890s. As a result, the subject contains terminology that might, on first reading, seem dry and unrelated to our desires to have a warm and mutually trusting relationship with our sentient horses. But it is through an understanding of the science of learning that we can really connect with our horses and be worthy of the trusting partnership that we so often crave (see also EBTA, 2012b; McGreevy, 2004).

Learning can be *associative* or *non-associative*. Non-associative learning takes place via repeated exposure to a stimulus (or "something happening"). It does not involve rewards or punishments, it is simply a response to a stimulus that is in the environment but cannot be controlled by the subject (in this case, the horse). *Habituation* takes place when the stimulus is a constant event that the horse gradually learns to ignore or, in conventional language, "he gets used to it". So, a horse in a field adjacent to a railway line may initially be frightened when a train passes. But over time, he will habituate to the presence of the train and his spook level and flight response will diminish. *Sensitisation* takes place when the horse becomes increasingly aroused in the presence of the stimulus. For example, a horse with an injury may become increasingly reluctant to allow a caregiver to treat him. Finally, *desensitisation* is the opposite process whereby the horse becomes less aroused by a process, for example a horse recovering from saddle-related pain and associated behavioural problems may learn first to accept a handkerchief laid on his back, building up to larger and heavier cloths or items, until the saddle can be placed across the horse's back without issue.

Associative learning is more widely and proactively used in horse training and involves the paired association between the stimulus and behavioural outcome. The famous "Pavlov's dogs" discovery – that the dogs would salivate prior to the ringing of a bell that signalled food – spawned the branch of learning known as *classical conditioning*. This is the process by which we subconsciously create pairings between two stimuli. Horses commonly learn to associate objects or experiences with fear: for example, a horse who has had a bad experience when wearing a head collar may develop an associated fear of the head collar. Careful training can reverse this process via *counter-conditioning*, whereby the stimulus (in this case, the head collar) is deliberately paired with positive experiences, such as eating high-value food, until the horse gradually starts to associate the stimulus (the head collar) with positive emotions. The counter-conditioning process is typically combined with systematic desensitisation, so that the desensitisation includes the development of positive emotions towards the stimulus rather than merely relaxed acceptance.

While classical conditioning plays a vital role in training, in that it governs the horse's emotional response towards the trainer and the training, it is largely the case that most training actively employs *operant conditioning* as the predominant way of achieving the behaviours in which we wish the horse to engage.

Operant conditioning is the process by which the animal learns that there is a consequence in response to a specific behaviour. If the consequence causes the animal to be more likely to repeat the behaviour in the future, then we say that the behaviour has been *reinforced*. If the behaviour is less likely to reoccur in the future then we say that the behaviour has been *punished*. If the stimulus leading to reinforcement or punishment has been added, then we say that we have used *positive reinforcement* or *positive punishment*. For example, a food reward in response to a desired behaviour is a widely used example of positive reinforcement and a smack for an undesired behaviour is a common positive punisher. If the stimulus leading to reinforcement or punishment has been removed, then we say that we have used *negative reinforcement* or *negative punishment*. For example, the release of rein pressure in response to the horse halting correctly is a common use of negative reinforcement. Negative punishment is less commonly used but a possible example might be removing yourself (i.e. attention) from the vicinity of your horse if he has started to mug you for treats or scratches.

It is important to understand that the degree of reinforcement or punishment does not change the definition; they can be mild or extreme. In particular, we often do not like to think that we punish our horses but, if even saying the word "oy" has the effect of reducing the likelihood that the horse will repeat a behaviour in the future, then that added verbal "oy" was still technically positive punishment, albeit a mild and debatably still ethical use of it.

Reinforcement is not as simple as merely continuing to provide a treat for every repetition of the behaviour. Such a *continuous reinforcement schedule* tends not to maintain the behaviour as the horse habituates to the training and/or reinforcers. Instead, it can be more effective to use more random production of rewards on a *variable schedule of reinforcement*, rewarding only selected presentations of the behaviour. The optimum rate of reinforcement required to maintain a horse's behaviour will depend on the individual horse and the skill of the trainer. Too many rewards might lead to habituation; too few and the horse will lose interest.

If the food reinforcer or punisher does not lead to a change in behaviour then it is just a *non-contingent stimulus* and operant conditioning has not taken place. This might be because the stimulus is insufficiently noticeable to the horse; we say the stimulus has low *saliency*, relative to the other stimuli present. We can give horses treats for "no reason"; some horses may simply eat the treats, whereas others will attempt to "mug", in which case the behaviour might become operant if the giving of treats coincides with the mugging behaviour. This has led to a common view that hand feeding encourages horses to bite but this is only true if the timing is poor. Non-contingent stimuli can also be aversive; we would all agree that gratuitous aversive stimuli are abusive and have no place in training at all. However, sadly, they are used all too commonly and trainers often unwittingly use them in the misguided belief that they are having a useful impact on the horse's behaviour.

Behaviours can also be changed via processes called *extinction* and *flooding*. Extinction is the phasing-out of a behaviour as the horse learns that there is no longer any consequence. Disconcertingly for novice trainers, the behaviour can sometimes worsen before it improves – for example, a horse who has learnt to mug pockets will not learn that mugging is ineffective until he has been permitted to mug for long enough for him to realise it is futile. Even once extinction has taken place, a long-held behaviour will always be vulnerable to *spontaneous recovery* of that behaviour, particularly if the behaviour is inadvertently reinforced. Extinction and spontaneous recovery are the results of a variable schedule of reinforcement, since the "withheld" reinforcers encourage a strengthening of the behaviour up until the point where the lack of reinforcement causes the horse to stop trying.

Extinction can take place as part of interactions with humans or merely as "part of life". Flooding, however, is a more controversial process that involves the horse being forced to confront a fearful situation until he no longer responds in a fearful manner. It is not a training method that trainers will admit to using; however, the rapid conversion of a frightened, non-loading horse into a horse who loads quickly during a trainer's demonstration, is a common example of popular use of flooding. Often the resulting demeanour of the horse is mistaken for relaxation, even happiness and willingness. However, the apparent compliance is accompanied by the subtle signs of stress, betraying the aversive and forceful methods that have led to the horse's state of *learnt helplessness*. The horse looks "shut-down", offering no behaviour without instruction, responding to cues without enthusiasm and gives the appearance of being depressed. Another common example of flooding is exposure of a horse to a plastic bag or tarpaulin. A horse's initial, fearful, response might be to jump around and spook, trying to avoid the frightening object. By employing undeniably effective techniques, a skilled handler would be able to insist on the horse standing still and prevent any attempts at escape. Eventually, the horse would be standing, apparently calmly, in the vicinity of the plastic. He may even permit himself to be draped in the plastic, still without responding in an overtly fearful manner. But this absence of a fear response does not mean the fear has disappeared; the fear is very much still present but is manifested differently, by standing still. The internal physiological signs and feelings of fear are still present. As the feelings of lack of escape intensify, the horse will gradually shut down, entering learnt helplessness, a state commonly confused with compliance. The description here sounds extreme, but it is so common within the equestrian world that it is something we have come to recognise erroneously as normal.

Reinforcing Jasper's behaviour

Jasper's case nicely illustrates the complex interplay between the clearly defined theoretical terms and the reality when we are faced with a practical situation in real life.

Initially, my stroking was making use of systematic desensitisation and my ongoing scratches were an intended source of pleasure that I hoped would counter-condition his experience to something that was better than just relaxation. That certainly worked to some extent. We also included non-contingent food to further help him associate hoof handling with positive experiences. As our work together continued, I was able to use the scratches more as a positive reinforcer that followed the correct behaviour, rather than an ongoing source of counter-conditioning. It was important for me to recognise that even my presence – and what I represented to him – was a potential source of "pressure"

Photo: Catherine Bell.

and that the removal of myself from his vicinity was a source of negative reinforcement. Pressure is not only physical contact – some physical pressure can be perceived as a neutral stimulus – and psychological pressure can arguably have a more significant impact on the horse than physical pressure. Obviously, I tried to keep this pressure to a minimum, attempting to avoid any increase in anxiety on his part. And of course, however much I wanted to avoid it, any accidental touch of his feathers had an inadvertent punishing effect and he would temporarily resist my presence near him, threatening to kick if I made a mistake.

What is the horse's experience of training methods?

With an improved understanding of learning theory, we are in a much stronger position to assess popular training methods and determine whether or not they are something in which we wish to participate.

Many trainers, whether deliberately or unwittingly, use a variety of euphemistic language to describe their approaches and techniques. Expressions such as "raising energy", "working in harmony", "developing a bond", "at liberty" or "force-free" can sound very appealing and the final result might look ethical but they do not always stand up to scrutiny. How were the behaviours trained? Did the horse demonstrate any of the small (or large) signs of stress in any stage of the training? Is there a consistency between what the trainer says and does, particularly when we attempt to adopt the perspective of the horse? Is the trainer recognising the elements of learning theory that are included in his/her training method?

What are the reinforcers? Are there any punishers? It can sometimes be difficult to tell whether a stimulus is punishing or reinforcing, particularly if coupled with the trainer's unscientific language. But the horse will "tell" us if we observe his behaviour carefully. We know from our definitions that a punished behaviour will diminish and reinforced behaviours increase. So, we can tell whether a horse finds the stimuli reinforcing or punishing – or rewarding or aversive – by looking to see whether there is a likelihood of the behaviour increasing or decreasing in the future.

Does it matter whether behaviour is being punished or reinforced? Is this just semantics over scientific jargon? Or are there ethical implications? I would argue the latter. We have already discussed what happens when a horse is caused some distress, with increases in certain hormones potentially leading to negative health consequences. If a horse is punished then he will feel a degree of fear, frustration and/or physical discomfort – we might not like to admit this, but the reality is that if the stimulus is not aversive in some way then it would be ineffective and the horse would not change his behaviour. What about reinforcement? We know from neurobiology that when something good happens, dopamine is released in our brains, particularly if the good thing is unexpected. Dopamine is a neurotransmitter that contributes to making us feel good, so much so that it is heavily implicated in addictive behaviours. But addiction aside, increased production of dopamine during training is going to add to the well-being of the horse. Conversely, the dopamine is reduced if there is also a build-up of CRF present, as in a stressful situation. Thus, the salience of the dopamine-triggering events compared with the aversiveness of the session will indeed influence the horse's experience and hence how ethically we are training him.

Traditional horse training and natural horsemanship

The problem for equestrianism is that almost all training – both traditional and natural horsemanship – makes use of negative reinforcement at best, and more commonly includes a lot of positive punishment as well.

Traditional horse training has always taken place via the application of pressure. We squeeze with our fingers, hands, seat or legs. If we are less refined then we kick, pull or yank. Sometimes we use whips or spurs. When riding we apply pressure in a variety of ways to the horse's nose, mouth, poll, sides, flanks and back. That pressure might be at a low level (or not) but its release nevertheless indicates to the horse the moment that he has performed in the desired way. This topic is addressed further in Chapter 4.

If the applied pressure is unsuccessful at getting the required response, most methods of horsemanship would ask again with more pressure. This reliance on pressure is partly due to the belief that if a horse is "disobeying" the instruction then he is being "silly" or "naughty". But as we have seen, the horse will have a valid reason for behaving as he does and this calls into question the notion of whether "naughty" is a valid concept in training at all. It merely lays the blame at the horse, whereas we should be taking on the responsibility for ensuring that the horse understands our requests, that our requests are reasonable (I would consider that many conventional requests of horses are not reasonable) and that we have provided sufficient incentive for him to engage with us.

Trainers may encourage their students to include pats or rubs, and these might indeed indicate to the horse that he has behaved appropriately. But they are not sources of positive reinforcement unless they have caused an increase in the behaviour; typically pats and rubs as part of a training system that is otherwise rooted in negative reinforcement will simply not be sufficiently motivating to have an effect. They are examples of stimuli that we define as being non-contingent, having no effect on the behaviour of a horse.

Natural horsemanship has become increasingly popular over the past 30 years and has led to several improvements in the lives of domestic horses. Proponents are more likely to make significant effort in meeting the ethological needs of horses by prioritising increased turnout and forage-based diets. They might avoid the use of bits and/or shoes, believing them to be detrimental to equine welfare. Above all, people following natural horsemanship training methods are likely to desire a relationship with their horse that is more rooted in trust than compliance, and to consider that whips, spurs and competitive sport are likely to detract from that close bond. As mentioned above, the language used by natural horsemanship trainers is often appealing but can be misguided, with the needs of horses sometimes not being met in favour of adherence to "principles". However, it is undeniable that the popularity of natural horsemanship has led to changes in the way we consider equine welfare and what it means to be a "horse person".

Despite the language of "trust" and "relationship", close perusal of the training processes lead us to observe that the training itself is rooted very firmly in negative reinforcement and positive punishment. The pressure is commonly applied via halters of very specific design. Some are "soft" rope halters but the narrow rope is able to transmit high pressures via the reduced surface area in contact with the horse's face. Some of these rope halters have knots located in sensitive parts of the face to enhance the pressure further. "Pressure halters" of various designs tighten when the lead rope is made taut, either when the trainer pulls the rope or when the horse attempts to flee the situation, and release when the horse complies. Some methods make use of round pens, with the horse being sent into flight until he learns that such behaviour is futile as there is no opportunity for escape. Other techniques make use of ropes being shaken and/or "thwacked" on the ground. None of these techniques take into account the reason why a horse may not be behaving in the desired manner but are merely achieving compliance through application of aversives or inducing fear.

Perhaps one of the most confusing things about natural horsemanship is that the end result after training can be very sensitive, including "liberty work", with the horse loose, inferring that he is happy to participate. For example, trainers may use very light pressure, perhaps even a flick of a fingertip in a certain direction, to ask a horse to move around them in a circle. However, this has usually been accomplished through earlier training that involved much stronger and aversive pressure. Over time, the horse learns that light requests accurately predict increased pressure, and so he responds promptly in anticipation. Thus, the training can be softened and refined. While this softening is obviously a good thing, the initial stronger pressure is counter to the shaping process discussed earlier and the advancement in behaviour is the result of fear and awareness of repercussions for non-compliance in earlier tasks – the horse has merely learnt how to behave so as to avoid pressure.

Similarly, a horse wearing a pressure halter may have the rope thwacked on the ground near him. This causes him to feel frightened and jump backwards, intending to flee. The halter tightens around his face and causes him pain, sufficient to worry him more than the fear of the thwacked rope. Here, we say that the fear of the pain is more salient than his fear of the rope.

Thus, the horse learns not to jump backwards when the rope is thwacked on the ground but to stay close to the trainer and with requests, even if those requests are painful, difficult or frightening to the horse. Critics believe this to be an example of flooding, and that the horse is forced to perform despite showing many signs of fear and stress. They do not consider the relationship to be based on trust but on fear and believe that the signs of supposed calmness and relaxation are signs of a progression towards learnt helplessness. When humans undergo similar experiences of feeling disempowered and helpless, they are diagnosed with mental health conditions including depression.

I am unapologetically one of these critics. Yet I acknowledge that there are many people who have been using natural horsemanship techniques for many years, without causing apparent distress to their horses. I asked a long-term proponent of natural horsemanship for her thoughts on what it is that appeals to people, and why they might prefer it to methods rooted in positive reinforcement (clicker training) that will be discussed in the next section:

People tend to be drawn towards the natural horsemanship approach because they see it as a kinder alternative to the conventional horsemanship they have been exposed to at livery yards, riding schools and similar. When asked why they favour it over clicker training, they might talk about their approach being related to how horses interact with each other naturally, saying that horses don't give each other treats, but they will reinforce their will on other horses physically. Imagine the scenario where someone has watched a horse in a clicker training demonstration frantically guessing what is expected of it, looking in vain to its trainer for a treat after each attempt until it finally stumbles across the answer. The stress on the horse is visible and it's understandable that the watcher might be of the opinion that it would be kinder to simply lift a lead rope, or apply pressure from fingers, to point the horse towards success.

Training with positive reinforcement

Personal bias aside, both behavioural psychology and neurobiology seem to be pointing towards positive reinforcement being the most ethical way to train a horse. And our own experiences often tell us this is true, since we tend to prefer being given a pleasurable reward than merely experience the removal of an aversive stimulus. Furthermore, although, technically, punishment can be effective as a training technique (provided the punisher is strong enough and aversive enough to cause the horse to change his behaviour), it is relatively inefficient since it gives the horse no information about the correct behaviour he should be performing. All it does is give the trainer a way of saying "don't do that". That is not very helpful for a horse who may be trying to do the right thing, doesn't understand what the right thing is and may also be in a frightening environment that is the reason for his non-compliance in the first place.

Horse training so rarely includes genuine positive reinforcement, let alone relies on it exclusively. Is it even possible? We know from the training of captive marine mammals that it is indeed possible to train a large mammal without resorting to physical pressure. This training takes place entirely via food rewards and therefore there is no reason why we might not train horses in a similar manner. I hasten to add that I am not endorsing the training or captivity of marine mammals, merely noting that the mechanics of training should be comparable.

The only popular method that tries to train with positive reinforcement, either alone or combined with pressure techniques, is clicker training. Clicker training uses a marker signal to indicate the desired behaviour at the precise moment it takes place. This marker signal is commonly a click sound made by a clicker, a plastic box containing a metal strip that bends with a clicking noise when pressed. Other signals sometimes used are verbal clicks, made by "clocking" the tongue or simply saying the word "good". Quiet clicks made by the mechanism of a ballpoint pen can be good for horses likely to find the clicker noise frightening, as is the opportunity to habituate or desensitise to the noise in advance of the first clicker training session. Choice of marker doesn't particularly matter; it is its classically conditioned pairing with a follow-up reward, either food or pleasurable scratch, that encourages the horse to repeat the behaviour again in the future. Thus, clicker training utilises both classical and operant conditioning.

Of course, the novice horse has no clue as to what he needs to do in order to obtain the reward. He may not even realise there is a reward to obtain! An introductory exercise that I have always favoured is to hold a non-scary but unfamiliar object close to the horse. Natural curiosity or accidental movement will normally ensure the horse touches the object. At the precise moment of the touch, the trainer should click the clicker and provide the reward. After a few repetitions, the horse will suddenly understand the association and start to proactively try to touch the target. Some horses take longer than others; horses who have experienced a lot of aversive training may find it particularly difficult to trust that offering a novel behaviour will not result in punishment. Please note that this brief summary of clicker training

Case Study: Talisman

Positive reinforcement can be applied without inclusion of the clicker; the reward is offered simultaneously with the behaviour instead of following the click. This example of working with a very young, semi-feral and largely unhandled pony describes the process and shows a nice contrast to Jasper's example above (see Figure 3.6).

I was standing in a field waiting for the owner to bring a horse to me for hoof trimming. In the meantime, Talisman appeared curious about me and started to sniff around me. She was loose and was free to leave at any time, both in terms of physical restraint and emotional conditioning (in that she had no reason to think that there may be a negative consequence of her walking away from me). I touched her back and, when she showed no fear response, started to give her a scratch. She immediately made it very clear that she enjoyed the scratches, with extended lip, soft eyes and repositioning her body in an attempt to elicit more scratches if I attempted to stop. Thus, in a few seconds I had trained the behaviour of "stand still and allow me to touch you". Think back to the criticisms of shaping and how it is perceived to be a slow process – maybe not!

Rather than risk reinforcing her for barging into me and becoming too demanding, I started to run my hand down her shoulder, in a similar manner that I had for Jasper but the key difference being that Talisman was actively requesting it; Jasper had been tolerating it. This level of proactivity was what made this case true positive reinforcement, rather than merely some form of manipulation on my part. I continued to give Talisman scratches in return

Figure 3.6: Talisman approaching me for scratches, but is happy to lift her foot in order to obtain them. Impromptu sessions such as this can be inappropriate, with muddy fields and other loose horses. But sometimes you just have to seize the moment and accept whatever the horse chooses to offer. Photos: Kelly Taylor-Saunders.

for strokes progressively further down her leg and, since she was so keen to continue, my biggest concern was overfacing her. I didn't want to risk exploiting her desire for scratches by lifting her legs before she was emotionally ready to do so. Instead, we switched legs and repeated a couple of times. By the time I reached her pastern I incorporated some shaping steps that included applying physical pressure to her leg in preparation for lifting it. Since her motivation for scratches was so high, I would argue that the physical pressure was not sufficiently salient to be perceived by her to be an aversive stimulus and instead was instantly counter-conditioned by the desirable scratches.

She continued to have no concerns about the physical pressure and therefore I continued and was able to lift each front foot, making no attempt to hold on. The first time she snatched her foot back quickly. Thereafter, she allowed me a brief moment of holding each front foot. She was still keen to continue and I was concerned about stopping suddenly in a manner that she may interpret as negative punishment. Instead, I was able to draw her over towards another horse and leave her while she was distracted.

When training is as free as this, then we can indeed consider it to be pure positive reinforcement. While with both Talisman and Jasper I was attempting to use positive reinforcement, Talisman was not feeling threatened in any way and so my presence was not a negative stimulus. Jasper clearly did not choose to be in my presence in the same way. This disparity between our intentions and the horse's actual experience of our training is a key problem for many horses, including, as we shall see, for horses being clicker-trained.

is not enough for a novice to "just get started" and I would always recommend learning via a registered equine behaviourist so as to avoid the many problems that can be caused due to insufficient understanding of the theory and practice involved.

It is also worth adding a note of caution that combining positive reinforcement with more aversive methods can be extremely damaging to the horse's welfare. When a horse is given the opportunity to offer a behaviour in return for a reward, he needs to feel safe to do so. He is placed in a very vulnerable position and may be very concerned that the behaviour he offers could be "wrong" with negative repercussions. If the trainer does indeed deliberately or inadvertently punish a response, then that fear of repercussions is reinforced and will be even more acute the next time.

When thoughtful training with positive reinforcement, with or without the clicker, is combined with the meeting of a horse's ethological needs, it is arguable that we are approaching the most ethical form of training. Horses will actively choose to participate, with demeanour very different from the apparent compliance that accompanies the harsher examples of pressure-based training. They can appear enthusiastic, sometimes unwilling to end a session. They can be confident in their decision-making; proactive participants in a training session. They do not fear the consequence of a wrong answer and will sometimes even appear to offer an incorrect response voluntarily in order to "check" their understanding. Jo Hughes is a behaviour consultant and trainer who has noticed a tremendous improvement in her horses:

Positive reinforcement training can contribute to an increase in equine welfare due to the mental stimulation and empowerment the horse gains from having the freedom to express their behaviour … Freedom to choose also has the wonderful side effect of developing really deep and trusting emotional bonds between humans and horses.

The example on the previous page illustrates how this freedom and enthusiasm can be manifested, even in relatively unhandled horses.

Positive or negative? Some thoughts from professional trainers

I wanted to add some perspectives from other trainers at this point. The problem is that most trainers are so rooted in commercial gain from their chosen methods that it is very hard to obtain objective views. Instead of asking representatives of the well-known "methods" I thought I would ask some trainers from whom we rarely hear, but spend their days working with some of the most traumatised and difficult horses, donkeys and mules.

Nikki Haddock is the manager of Ferne Animal Sanctuary, a centre that is responsible for the care of over 300 domestic and farm animals, some of whom are horses and donkeys. I liked Nikki's ethological approach to training, allowing horses the opportunity to integrate with the herd before any normal training started:

My role in training is to help guide the guys who work with me to work on the less challenging animals, the more challenging ones are up to me to work with and manage until they are ready to find homes. I go about this initially by integrating a new animal with the herd and observing their normal behaviour in their own environment with their own kind. I wouldn't consider starting training for, at the very least, two weeks. I like the animals to be comfortable and relaxed in their own environment before I start asking anything of them. I do very much feel my way (I do have shaping plans in mind should I need), I am happy for a training session to be guided by the animal if the animal is offering me some behaviour that I might want or be planning to work on in the future.

The schedule I work by is the horse's schedule; it takes as long as it takes for the individual as long as some progress is being made. I like to use scratches or food rewards for desired behaviours being offered (this is the only time lice are an advantage on a new animal! You can quickly become their buddy when they are so itchy!).

We recently had a pair of ponies in who were terrible loaders. We spent around three weeks using food rewards (they were little natives who also had sweet itch so food and scratches were super high value to both of them) with them when they interacted with the horsebox. One was self-loading within a week; the other was more timid in nature so he naturally took longer to trust that this was a good idea. At one point we had one at the stage of going straight in the truck for a few treats while the other was getting a treat for having his front feet on the ramp for five seconds! This worked very well for these guys so I would not have changed anything.

We did have a little gypsy cob who was quite angry at being handled, she did however have lice when she came to us so it was easy to use positive reinforcement with her but she had a very short attention span and would regularly throw her weight around so when training standing still, as soon as she started fidgeting the scratches would stop. This combination of giving scratches and taking away scratches was very effective for this pony. I don't feel this put undue stress on the horse. She was in foal so we had to get her handleable in a short time frame. Had we had more time I would have probably made the standing still time shorter and ended the session before she would get fidgety. At this kind of level I do not have a problem with using negative punishment. I like to use positive reinforcement as much as possible but sometimes I need other options which do not cause harm to the horse.

Lisa Lanfear is a welfare officer for Communities For Horses (CFH), a charity that provides outreach services to the urban horse-owning communities. According to the CFH website:

The urban horse population in Swansea comprises an estimated 600 horses and ponies owned by an estimated 300 members of the community. The urban horse owners represent some of the most deprived people in the UK, the majority of horse owners are under 25 years old and suffering great hardship, often abuse. We have concerns for the welfare of the urban horses in Swansea, not only those that are tethered but also those that roam common land and marshes and those kept in gardens, sheds and stables.

This gives Lisa perhaps a unique outlook on training, having to make compromises between the needs of the horses and owners. It is crucial to shape the attitudes of the humans and however much one might want to insist that training utilises positive reinforcement, the realities are that more improvements can be made for the horses if a more pragmatic view is taken.

Lisa explains:

The urban horses in general only know aversive methods of training – flooding and water deprivation are common, as is beating and I also come across some extreme and unusual methods of training. Initially when working with the owners I tend to use pressure and release, to show the horse, and owner that training can be different to what they are used to. Also, this way I can gauge what, if any, understanding the horse has regarding some typical "asks" and what areas need to be worked on or developed. I visualise a shaping plan as a flow chart, using a stepped approach when working with both children and adults for their issues with ponies, getting them to think outside the box and about the process as well as the result they are aiming for.

I often use a combination of gentle pressure and release and positive reinforcement, when, for example, training a horse to be led I would be more likely to use just positive reinforcement with my own private horses but with rescues, and horses in the communities, who tend to bolt and where time is limited I find that such a combination works best. Also, the owners are more likely to adopt this approach themselves than if I promoted positive reinforcement alone – we have to shape their behaviour too! When using clicker training, I usually use a target for the horse to follow and phase it out when the horse has learnt the concept of leading nicely. However, when a task has been performed it's nice to say "thank you", and mine still, to this day, get a treat after having all their feet picked out or walking to the field nicely. Not a walking vending machine, but just reinforcing that good behaviour gets rewarded and that I'm nice to be around!

I think positive reinforcement used in conjunction with gentle pressure allows the trainer to see that gentle approaches do work, and you don't get rope-burns or kicks. Working this way allows people to grow with the horse they are working with. Although it might take a longer period of time, small steps make great journeys.

I think for the professional "average Joe", who turns out riding horses in six weeks, and note that some urban horses are broken to ride and drive in a week, it is going to be a challenge to change their behaviour because there is a cost implication to taking things more slowly. However, this is why we are focused on long-term outreach, change will happen little by little.

Maisie Wake is a behaviourist who was fortunate enough to work with the mules whilst employed by the Donkey Sanctuary. She is now the main equine behaviourist at Munchkins Miniature Shetland Rescue Centre. She reflects on her approach to working with rescue cases:

> I remain extremely methodical and use either scratches or food for positive reinforcement. I tend to use a variable schedule of reinforcement where possible and find this extremely effective for equines. If I am teaching a new task however, I may start with a continuous schedule of reinforcement and a very high rate of reinforcement. I quickly shift this to a variable schedule of reinforcement, but the rate of reinforcement may remain quite high. As the equine becomes familiar with what I am asking, I will also phase out the clicker if I am using a clicker, and begin to use a voice signal as a bridge instead. I eventually phase out the rate of reinforcement as well, but keep some level of reinforcement ... I do everything I can to avoid frustration during a training session, and try to ensure that they have water, friends, and other food sources available to them during a training session.
>
> I have often taught volunteers how to safely lead ponies using negative reinforcement. It was a common sight to see a handler tugging on the pony's lead rope and the pony standing stock still. The handler would then walk back towards the pony and give him/her some fuss, then repeat the futile tug of war. I have successfully pointed out that they are actually reinforcing the standing still behaviour, by releasing pressure at this point, as well as going to the pony and giving him and her some fuss – so providing a combination of negative and positive reinforcement. I have taught them to instead remain consistent when asking to move forwards, by the gentle pressure on the lead rope. I explain that this should not escalate, but instead remain the same, and be instantly removed when the pony moves forwards. Although it may be far nicer for the ponies to be taught to lead through positive reinforcement, I have found this very difficult to put into practice in an environment such as a horse charity. Many of the ponies already have an understanding of negative reinforcement training, and in a gentle form, it is clear and easy for them to understand. The same goes for the handlers.

Is pure positive reinforcement our goal? What would horses choose?

It certainly sounds as though I have been recommending pure positive reinforcement as a goal and I would agree that there are many cases – especially horses who have had traumatic experiences and/or are particularly fearful – where to use anything other than approaching as close to pure positive reinforcement as possible would be tantamount to abuse. A semi-feral pony such as Talisman was a good example. So why would I hesitate to recommend it?

Whether or not I recommend training with positive reinforcement, and particularly clicker training, depends strongly on how much training takes place and in what context. Positive reinforcement elicits that strong dopamine spike particularly when the reward is unexpected. If the training takes place repeatedly and for long sessions, then there is nothing unexpected about the rewards in those sessions. The horse tends to habituate to the reward, as we saw for continuous reinforcement schedules. In many cases I have seen, there doesn't seem to be any behavioural indication that the horse is finding the session pleasurable. Sometimes horses in such situations seem frustrated or detached; some geldings demonstrate sexual arousal that is more indicative of something being wrong than the more common joke that "he's having a good time". Proponents of clicker training will claim, in response to articulation of these concerns, that these problems are caused by inexperienced trainers. But this is not always the case; indeed, sometimes the problems are worsened by experienced practitioners, who try to fix problems caused by clicker training, with yet more clicker training.

The problems caused by clicker training need to be solved by stopping the clicker training, at least in the short term and possibly for good, and then addressing the causes of those problems. The causes are not always clear. Sometimes they are fundamental environmental/ethological concerns. From various discussions I have had with other behaviourists, other suggestions include addiction to the dopamine spikes and/or boredom and frustration with the training. Some horses seem to resent being cued to perform particular behaviours and struggle with the degree of control required of them. After all, the nature of clicker training is to reward the horse for offering a behaviour; if he chooses not to offer the behaviour and not to obtain the reward then that seems to me to be a reasonable part of the transaction. If we are "showing the horse who's boss" and insisting on the behaviour then, treats or no treats, I would argue that our training is as authoritarian as the traditional training that so many of us chose to leave behind. Clicker training is often described as "force-free training" and while that certainly can be the case, it is not guaranteed.

The debates surrounding the ethics of positive versus negative reinforcement often forget one thing. What would the horse choose? (See Figure 3.7.) If pressure is required in order to make a horse comply then it is probably clear that the horse has no choice. "Do as I say or I'll hurt/scare you" is an unenviable "choice". But what about choice in the context of positive reinforcement? Choice to participate was a key difference between Jasper and Talisman's experiences, although my techniques were largely similar and even Jasper's training was rooted firmly in positive reinforcement. The problem was that he did not find the rewards as salient as Talisman, relative to the training. Ethical training requires us to make those judgement calls as to when it is acceptable to insist that the horse complies, and when we can allow the horse to walk away.

Maisie Wake again had some valid points to

Figure 3.7: Choosing to investigate a novel stimulus. Photo: Catherine Bell.

make about the balance between the different elements of operant conditioning:

> While I feel that we could radically increase our use of pure positive reinforcement, it is valuable to be aware of all four quadrants of learning, and have a firm understanding on what to avoid as well as what to use more of. The lesser used negative punishment has its place in teaching a horse some rules around grooming. Some people are very reinforced themselves by the act of mutually grooming with their horse, and it can also be pleasant for the horse. I have successfully guided these people through the process of immediately removing the positive stimulus (them giving the horse a hard scratch) when the horse's grooming technique becomes too vigorous for human skin. If it is preferred that the horse wiggles their lips in the air and doesn't make contact at all, this can be shaped through a process of positive reinforcment (adding the grooming again) combined with negative punishment (stopping the grooming if they start to nuzzle). I have found equines learn this so extremely quickly, that it has helped families with young children especially, to build an environment where the equine can still get loads of scratches but doesn't start to get too rough for the young handlers involved.

In a world where you can analyse your horse's day-to-day experiences (with or without a human handler involved) into the four quadrants, I personally think that it may complicate things to become fixated on pure positive reinforcement, or perhaps even fearful of negative reinforcement being used. At the end of the day we need to enjoy our experiences with horses, whilst ensuring that it is not at the horse's expense. Ideally, there should be something that the horses can gain through our interaction as well, so positive reinforcement should definitely come into our interactions in some way, shape or form. For a healthy relationship with your horse, I feel this should be varied and fluid. It is important to sometimes just "be" with your horse, identifying what they enjoy most, and bringing that into your interactions in some way. Becoming set on positive reinforcement training risks bringing an element of obligation into the interactions, which could reduce the horse's ability to act out of autonomy and individuality.

The beauty of shaping is that it means that the choice of whether to use positive or negative reinforcement becomes less significant. The steps are sufficiently small and the horse's concerns are sufficiently considered that any pressure used can remain minimal. But this should not be misinterpreted to mean that positive and negative reinforcement are equivalent. I would still recommend that positive reinforcement is the predominant form of training, any use of physical pressure is counter-conditioned, even tiny signs of stress are averted and any use of pressure, physical or emotional, is both minimised and justified. I would also request that the horse's choice and autonomy are prioritised firmly above any agenda we may have for our horses. If we can do this, then we combine the science of learning with the bonding, trustful relationships so desired by proponents of natural horsemanship.

Some controversial points to ponder

The nature of ethical training is an extensive subject and everyone will have different ideas as to what is acceptable and what is not. And our ideas will evolve over time. These ideas will vary according to the individual horse and his experiences as well. So, I think we should continue to reflect on them indefinitely.

What should we do if a horse has experienced extremely aversive training but has learnt to avoid the punishment and just "shuts down", blindly obeying any instruction? We might desire to rescue the horse and give him a retirement of luxury, but some horses might find the inactivity and lack of direction extremely stressful. Maybe their "shut down" façade helps to keep them feeling safe and, if they can predict and avoid aversive handling, should we remain concerned? Do we really want to risk opening them up and do we have the skills to support a horse who switches quickly from "shut down" to "opinionated"? Do owners really want opinionated horses who will sometimes say "no"? "Opening up" a horse leaves him very vulnerable emotionally – is it sometimes better not to try? Does it really matter what method people use, as long as their training includes shaping? Shaping allows us to keep aversive stimuli to a minimum so is this the best way forwards for "traditional" equestrians? Would horses be better off if all equestrians suddenly adopted positive reinforcement methods? Are horses better off having experienced and competent, yet aversive, training or switching to novice, clumsy attempts at clicker training?

These, and many other, unanswerable questions require us to continue thinking about behaviour. Training should not be merely about the horse as a passive recipient of our attempts to create or reduce behaviour. When we keep behaviour in mind we go beyond this; we see the needs of the individual horse, we use shaping and we walk the fine line that is our complex, balanced horse–human relationship, where both partners acknowledge and respect each other's autonomy.

Final thoughts

Ultimately, maybe the most ethical trainers are simply those who know when to stop. They understand that just because they *can* train behaviour, doesn't mean that they *should*. Consequently, the repertoire of behaviours they can train may remain less extensive than that of trainers who are more focused on achieving behaviours using whatever means possible. We will all disagree on what we should train, whether we should ride horses, whether aversive training is justified and for what goals – and even whether we should train at all. We will no doubt change our minds as we go along, sometimes increasing and sometimes decreasing the amount of training in which we choose to engage. The most important thing is that we consider the needs and experience of the horse at all times, keeping his behaviour in mind in all the requests that we make and giving him the opportunity to choose.

References

EBTA (2012a) The Ladder of Fear. Available at http://www.ebta.co.uk/lof.html (accessed online 26th February 2018).

EBTA (2012b) What Behavioral Science Do I Need To Know? Available at http://www.ebta.co.uk/faq-definitions.html (accessed online 26th February 2018).

Horseman, S.V., Whay, H.R., Mullan, S., Knowles, T.G., Barr, A. and Buller, H. (2016) Horses in Our Hands. Available at: http://www.worldhorsewelfare.org/survey-equine-welfare-england-and-wales (accessed online 26th February 2018).

McGreevy, P. (2004) *Equine Behavior: A guide for veterinarians and equine scientists*. Saunders, Elsevier, Edinburgh, UK.

Pryor, K. (1984) *Don't Shoot the Dog*. Bantam Books, New York.

Sapolsky, R.M. (1994) *Why Zebras Don't Get Ulcers*. St. Martin's Griffin, New York.

Chapter 4

Equestrianism

Debbie Busby

Equestrianism refers to a range of skilled activities performed by horse and rider or driver. The most well-known equestrian activities are the competitive sports of dressage, show jumping and eventing, or horse trials. Other equestrian activities include hacking and trekking, or the faster trail riding, endurance, horse racing, driving, hunting and polo.

In this chapter, I will consider how various aspects of equestrianism represent strengths or challenges from a behavioural perspective, and I will explore how specific equestrian disciplines affect horses, and how the riders and drivers who practise these activities can engage in them in a more behaviourally-minded way.

Tom the riding horse

Tom is a 5-year-old, 15.1-hh, chestnut thoroughbred cross who was brought to England from Ireland by a dealer nine months ago. In England, Tom lived outside permanently with a group of other horses on 60 acres of open farmland, which included copses and a river running through an undulating valley. Tom had been backed and started (introduced to a saddle, bridle and rider, and taught the basic go, stop and turn aids) in Ireland, and once the busy dealer had confirmed for herself that Tom knew these cues, she left him to run with the herd, checking him daily by Land Rover.

Tom was bought by Denise in the autumn after she saw him advertised in the local newspaper, and he was Denise's first horse. Denise had ridden for over 10 years when she bought Tom, first on a friend's pony and later having a weekend job caring for, exercising and competing a riding club horse who was stabled nearby. Denise had visited the local horse trials where she was overawed by the fit, obedient horses and their skilled riders, successfully negotiating the variety of imposing and "rider-frightening" obstacles on the cross-country course. She even saw Princess Anne take part, with no royal coterie in tow, mingling with the other competitors and driving her own horsebox away from the event. Denise had been thinking for some time about buying her own horse and this experience decided her. And that weekend she saw Tom advertised in the local newspaper. Denise went to view Tom and the dealer drove her down to the valley to catch him. It was early winter by now and the horses soon appeared at the side of the Land Rover because they knew that it delivered hay every day. The dealer had no saddle to sell with Tom and Denise accepted her invitation to ride Tom the half mile back up to the yard in just a bridle, as a way of trying him out. By the time they arrived at the yard, Denise had fallen in love with Tom and she agreed a price with the dealer. She arranged stabling locally; because it was winter, Tom would have to stay in as there was no winter grazing on this yard, but Denise thought that would be OK as she would be able to exercise him every day. She looked forward to riding Tom out and about on her own around the fields and the lanes, as she had done on the riding club horse she had previously ridden. But things didn't go according to plan for Denise.

The first thing she noticed when she approached Tom's stable was that he appeared about a hand bigger than when she had bought him. She could see his head over the stable door, high, tense and with ears fixed forwards. When she went in to groom him and tack him up he wouldn't stand still and kept circling the box in an agitated way. She managed to get his saddle and bridle on and mount him, and he marched out of the yard, not listening to Denise's aids for him to slow down. Unfortunately, Denise didn't enjoy her ride as much as she had hoped because Tom would not slow down and when she cantered, she felt that she was being run away with. Things went from bad to worse and each time she rode him, Tom felt stronger and less controllable, almost bolting, Denise felt. He had also developed a new behaviour of refusing to go forwards when they reached a certain distance from the yard, and when Denise applied more leg, he would rear up. On the next three rides, the rearing became worse and on two occasions, Tom reared so high he fell over backwards with Denise underneath him. It was at this point that Denise realised she was out of her depth and she phoned the owner of her previous ride, the calm and well-behaved riding club horse, to ask her for help.

In the rest of this chapter, we will look at the factors involved in a riding horse's life that might have contributed to Tom's problems, the advice Denise's friend gave her and the positive effect that this had on Tom's behaviour. As well as Tom's story as, at this stage, a not-so-happy hacker, we will explore specific equestrian issues and disciplines to see where their behavioural strengths and weaknesses lie and how the riders and drivers who practise them can engage in these activities in a more behaviourally-minded way.

When we want to apply the findings of behavioural science to the way we train and manage our horses, it is useful to know what a horse's natural needs are. More and more research studies are finding that allowing horses to express their innate, species-specific behaviours on a day-to-day basis makes for improved well-being and a happier, healthier equine companion or athlete (McGreevy, 2013). The equid ethogram and behavioural needs of the horse, together with an overview of how equine welfare might be assessed is given in previous chapters.

Behaviourally-minded equestrianism

There are certain behavioural needs that are very important to consider when we take part in equestrian activities, whether they take us off the yard or even if our horses remain at home. Some horses travel long distances to events, even flying round the world to major competitions, and this can take its toll on their well-being. Over-stabling and not allowing sufficient turnout can have a similar effect, as can a lot of the beautification and preparation that we think our horses should undergo in order to meet our competitive objectives! How can we address these aspects of equestrianism in a more behaviourally-minded way?

The things we like to do with our horses can meet many of these basic needs and can therefore contribute to their contentment and well-being. For example, the fact that we too are a social species, and we like to go for rides, or go to shows with friends, means that we are giving our horses the opportunity for essential exercise, at all paces and in company. Horses need to walk, trot, canter and gallop to keep all their physiological systems in good working order, and many of our riding activities provide opportunities to do these things. Horses don't like to be alone, so if we can ride out, school or go to events in company with other horses, especially ones who are familiar to our horse, this will help them to remain calm and enjoy the experience. Riding out regularly with other horses from our fields or yards and returning back to the same place creates a mental map for the horse of a wider home range and matches their species-specific need for exploration of this familiar area. If we take them to unfamiliar places such as new events or a new trail ride, we should allow them to explore these new surroundings with our reassurance and some nice neck scratches or tasty treats, and all the better if we can take them to new places with familiar companions who will help them to feel more secure. We are also taking care of their body maintenance during our daily visits by performing regular health checks, and when we groom we are enhancing physiological systems such as circulation and developing muscle tone.

As well as allowing our horses' social needs to be met by taking them out with their field mates, we too can be a social resource for this affiliative species. Horses can develop close relationships with us in the same way that they bond with each other; this is another way in which equestrian activities can keep behaviour in mind. Anyone who has heard their horse's welcoming nicker when they walk into the field or stable knows what deep connections we can have with our horses. We can make great use of this bond to enable other basic needs to be met, including exploring with a friend, and playing. Taking our horses out for exploratory walks around the yard, field or tracks on a long line is a good way to let this happen, and gives them opportunities for oral object play, which is also something many of us have seen our horses do with stable door bolts and lead rope knots when tied up, too!

Travelling

Travelling can be very stressful for horses; asking a prey animal who relies for safety on long distance vision and wide-open spaces to enter a small, dark, enclosed box almost defies expectation, and habituation through sympathetic training over several months is required to allow a horse to get completely used to this experience. In addition, horses do not feel safe alone and this is an added stressor which we can easily avoid by always making sure that our horse travels with a friend. Having a friend with them will also help them to settle more easily into the unfamiliar environment at the other end of the journey.

Grand Prix dressage rider and British team trainer Paula Cooke BHSI says:

> I always sing the praises of my little travelling pony to anyone with a horse who gets anxious when travelling. Fiver (his nickname because he cost me five whole pounds) travels with everybody, everywhere, I use him to keep horses company on journeys and it's great if I have two competing as it means the one who is left in the lorry isn't on its own and doesn't get stressed.

It is also important to consider the direction that horses face when they are travelling; whilst we typically transport them facing forwards or sideways at an angle, the most robust studies (Cregier, 2009) have shown that they prefer to face backwards; this is thought to be because they can distribute their weight more easily and brace themselves against gear changes and braking.

Travel can have an unhelpful effect on horses' sleep patterns too; in order to understand why this might be, we can look at what is normal resting and sleeping behaviour for horses. Their resting and sleeping behaviour is divided into standing resting (called loafing or idling), standing sleeping and lying down sleeping, either supported on their chest (sternal recumbency) or flat out in lateral recumbency. When grazing time increases, resting time decreases, and this changes according to the seasons, but time spent lying down remains consistent and occupies around one hour a day. Unlike humans, adult horses are crepuscular (most active at dawn and dusk) with polyphasic sleep patterns; sleeping, idling, dozing or resting for around three to five hours per day, in short periods of time throughout a 24-hour cycle, and mostly at night. In groups or pairs of horses who are familiar with each other, individuals take it in turns to lie down and sleep, so that there is always at least one horse who remains standing alert for any threats (Mills and McDonnell, 2005).

Horses lie flat out in lateral recumbency and enter rapid eye movement (REM) sleep for about 30 minutes each day in bouts of three to 10 minutes, during which time it is thought that they dream because they sometimes vocalise or kick out and make running motions. This REM phase is essential to biological and psychological well-being and if they cannot lie flat out because they are anxious on their own, or because they are travelling and standing over several nights in a confined space such as a trailer, transporter or small temporary stabling, they can quickly start to suffer from sleep deprivation and have been known to collapse because their bodies still try to enter REM sleep whilst standing up and lose muscle tone. This effect of sleep deprivation is often incorrectly thought to be narcolepsy (Bertone, 2015). Researchers have found that when horses are transported, their REM sleep is disrupted for up to three days following travelling (Coumbe, 2001, cited in Fraser, 2010) so this is a significant risk to their behavioural well-being, particularly if they have other causes of stress and anxiety such as being in an unfamiliar location, new smells, different bedding or flooring, or significant temperature changes.

We can minimise these risk effects by being more thoughtful about how often we travel our horses, and the conditions in which we transport them. Reducing as many potential stressors as possible avoids a build-up of anxiety, and simple measures like having a trailer roof that allows light through, using familiar bedding in the trailer and adding some of the horse's own bedding from their stable, as well as always taking a travelling companion, allowing sufficient rest and recovery time on arrival, and having temporary stabling big enough to lie down in, will all be of benefit and allow both horse and human to enjoy the activity that awaits at the end of the journey.

Flying is a particularly specialist form of transport, which is used where valuable horses are moved around the world, for competition, breeding or sales purposes. It's intuitive to think that the noise and temperature conditions at airports and on planes might upset sensitive horses; however, surprisingly few problems are reported and the grooms who travel with their charges take extra care to make sure they are calm and relaxed before, during and after the flight (Horse & Hound, 2014). Horses are reported to sleep for longer periods during and after flights, however, and, given what we know about sleep, this might be beneficial. There are a wide range of factors to take into account; including ensuring horses are fed and well hydrated during the journey, and dealing with quarantine regulations in different countries, for example when horses are flown to Australia they have to spend two weeks in quarantine in the UK before they leave. This is because Australia has no equine flu virus, so their horses are not vaccinated against it. Blood tests, nasal swabs and temperatures are taken before flying and must all be clear, to make sure the horses are in good health, and all horses are quarantined as soon as they arrive abroad and for the duration of their stay. In America, horses are kept in isolation for at least 72 hours, which could have a significant behavioural effect; it would be good practice in this situation to make sure their groom is someone familiar whom they trust, and this is usually the case.

Nowadays, horses are usually flown in custom-built aircraft adapted specifically for transporting horses, and for air safety reasons nobody is allowed to stay in the hold with the horses during take-off. Behaviourally, this does not appear to present a problem to the type of horses who are flown by air as they are usually used to travelling in horseboxes, a similar experience. The specially built air transport boxes even allow horses to be untied, move around and stand forwards or backwards as they prefer, which in some respects may have advantages over typical road transport.

Some horses are given hydration drips before they fly to ensure they absorb the right amount of vitamins and minerals, and electrolytes can be given during the flight if dehydration is suspected. There is always a vet on the flight, who can administer intravenous fluids. It's crucial to avoid colic in this situation so feeding follows the good practice of little and often, and some grooms like to feed off the floor, so any effects of the plane's air conditioning system can drain out.

After landing, horses' temperatures are checked and they are rested for 36 hours; during this time, grooms lead them out in hand to graze, which helps them to relax and settle. With racehorses, all the race conditioning work is done before the horses fly and they are just kept ticking over in the days before the race. The grooms who fly with horses report that the horses recover from jet lag quicker than they do, and we might understand this from a behavioural perspective because we know that horses are polyphasic sleepers, unlike humans who are biphasic. This ability to sleep in several short sections within a 24-hour period is probably what helps them adjust more easily to different time zones.

Preparing for competition

More than ever before, we expect our horses to accept all sorts of special treatments in preparation for the big day, whether that's clipping, bathing, or mane and tail pulling. None of it is essential maintenance in the same way that horses' natural behaviours, like mutual grooming or rolling, are and most owners will know that our horses aren't always that keen on having this kind of pampering carried out. In fact, mane and tail pulling are actually painful, so who wouldn't object to that! There are many tools available now for thinning manes to create a "pulled" (thinned and shortened) appearance, and the point of pulling a tail is to show off the hindquarters, which can be done just as easily by plaiting the tail; this looks far more attractive than short hairs sticking out of the top of the tail, or the "shaved" look that some people use as a quick and easy solution. Allowing your horse to keep all his own tail hair allows him to put it to good use as extra cover in windy weather when he positions himself with his back to the wind.

Neither is it at all essential to bath our

horses regularly, unless there is a medical need. It strips the protective oils from their coat and leaves it dull-looking, and it's never a pleasant experience for the horse, so it's easy to flood them, psychologically, which can have associated unwanted behavioural effects as mentioned elsewhere in this book. It is far easier for them to remove their own sweat by having a good roll in the field or sand paddock after a ride or a schooling session. Clipping is another procedure that horses don't always take to very kindly and risks flooding them. It's a straightforward procedure to habituate a horse to clippers using a shaping programme including things such as an electric toothbrush to mimic the low-level humming sound, and miniature manual clippers to accustom them to the feel of the clipping effect, and we can incorporate this kind of husbandry training into our day-to-day activities with our horses. Clearly, clipped horses will need some other way to keep warm in the form of rugs, but if your horse isn't clipped, do consider whether you need to rug them at all. Leading equine physiologist Dr David Marlin says:

> due to its size the horse does not lose heat as rapidly as we do. So at the same temperature the horse will actually feel warmer than we do! Horses are incredibly adaptable when it comes to climate and are found in both some of the hottest and the coldest places on earth ranging from -40 to 60°C.

We can often imagine that how warm or cold we feel is a good guide to whether we should rug our horses; but as Dr Marlin explains, "No, we lose heat more rapidly than horses. We will feel cold when horses still feel comfortable. Our thermo neutral zone (when naked) is 25 to 30°C – much higher and much narrower than that of the horse." He also advises an hour a day with rugs off so that the horse's body can absorb essential Vitamin D, and reminds us that wind, cold and rain will induce the greatest heat loss, so natural or artificial shelter is very important (Marlin, 2017).

We also use more equipment to help our horses stay in good physical shape too, like solariums, hydrotherapy and horse walkers. These are very useful additions to a yard or barn when used appropriately and correctly but remember to always habituate your horse to these unfamiliar experiences, and don't overuse them unnecessarily. For example, there's no need to bath your horse every week just because you can dry him off quickly in a solarium; however, solariums are great for easing tired muscles. Likewise, don't let the horse walker take the place of a good romp in the field, or hydrotherapy replace a ridden exercise session without any good reason, but do make good use of nature's own hydrotherapy by letting your horse enjoy paddling, splashing and nosing around in safe ponds, streams or rivers while you are out and about with them, or provide this for them in their field so they can have the same fun with their friends.

Behaviour in equestrian disciplines

Dressage

Horses are embodied beings in a way that we humans find difficult to conceive. It is through their physical movement that they escape from danger and find food, water and other resources essential to survival and good health. Horses love to move and offering them that opportunity through dressage (which literally translated means "training") is a source of cognitive and physical enrichment as they learn to follow the cues given to them by a sensitive and sympathetic rider. We can use dressage to enhance our horses' day-to-day lives as we give them manageable mental and physical puzzles to solve in the riding arena or indeed through in-hand training.

Danish trainer Kirsten Alexa Hansen understands a horse's natural instinct to move in balance. Educated in the classical Austro-Hungarian riding system under Colonel L. Tersztyanszky, Kirsten concentrates on rehabilitation and says:

> My task is to "refresh" the horse's natural movement, which can only be established if you do not induce new tensions. Of course, it is important to create basic trust. The horses become very calm and interested in the work when you allow them to move in balance by using their natural functions.

Dressage involves good thinking skills on the part of both horse and rider and we can help this along if we provide daily turnout in a setting where they have company, foraging opportunities and places and objects to explore. Studies show that animals who live in enrichment environments have increased brain activity and are better at learning and problem-solving. It's a good idea to cross-train your dressage horse with some easy jumping or hacking out in between flatwork training sessions. Renowned Classical Dressage trainer Sylvia Loch makes sure that her horses are turned out every day and at 21 years of age her Lusitano stallion Prazer is still very clearly enjoying his work. Sylvia says, "none of my horses work more than one hour a day with some cross-country work in between and at least one day off per week." Sylvia is also a great proponent of shaping and reward-based training; she says, "the only way we can tell a horse that he has done something correctly is to reward him", and, "the training path of every horse should be done step by step – systematically and logically". A good working behavioural understanding of how to shape new movements and how to reinforce the behaviour we want will pay dividends. Sylvia is also keen to encourage a gradual training programme: "Training slowly in this way builds muscles, mobilises joints and gives the horse time to develop physically and mentally. It also builds trust".

A clear understanding of psychological learning theory (Baldwin and Baldwin, 2001) will also avoid issues such as horses becoming unresponsive to the aids or confused by what we are asking them to do. BHS Fellow, Assistant Professor and Member of the Council of the International Society for Equitation Science, Angelo Telatin recounts a time where a student was having difficulty getting her horse to canter: Angelo spotted that the rider exaggerated her hip movement just before she applied her leg aid to canter, and her horse was actually responding to her hip movement; he had learned this as his cue to canter, rather

Photo: Jenni Nellist.

than her leg aid. Angelo demonstrated this to the rider by asking her to exaggerate her hip movement a little more, and the horse performed a beautiful canter depart, showing that this was the cue that was under stimulus control, as far as the horse was concerned! He explained that she would now need to retrain her horse to canter from the correct leg aid. This misunderstanding of which cue the horse is actually responding to is something that a lot of riders do and is a great example of how miscommunication between horse and rider can cause training problems.

Jumping

We ask horses to jump obstacles in a variety of disciplines including show jumping, cross-country, working hunter showing classes, steeplechasing, point-to-point racing and hunting. If they are trained appropriately, horses can learn to enjoy jumping, and it's another opportunity for bodily movement and muscular development. We need to make sure our horses are physically sound, too. As with dressage, knowledge of how horses learn (McGreevy and McLean, 2007) is useful so that we avoid overfacing and flooding horses, causing problems such as running out, refusals or rushing fences. Using excessive force to drive horses at fences they are overfaced by or using too much tack to control them when they rush through fear, pain or anxiety, is counterproductive and leads to a horse becoming less and less keen to even enter a jumping arena. Gradual habituation to different types and heights of fences is the way to avoid these issues, and training horses to jump in a systematic way using the behavioural principles of successive approximation, or shaping, reinforcement of the behaviour we want, and creating positive classical associations of the enjoyment of movement, problem-solving and treats or scratches, will result in a horse who knows what to do and takes enjoyment in doing it.

Specific jumping disciplines require habituation to different elements of the overall experience, for example at showjumping competitions there can be a lot of unfamiliar horses warming up very close to each other; this can cause unhelpful anxiety that will adversely affect your horse's performance, so think of other options such as doing practice fences at a quieter time and keeping muscles warmed up by working in a quieter place; leave entering the collecting ring until later, a few minutes before you enter the ring. A young or novice horse can be affected by your competition nerves too, so think about reducing these by including "clear round" or *hors concours* (literally, "outside the competition") jumping where the competitive element is eliminated. This is a great way to let your horse see the sights and sound of the showground or the indoor arena without having the pressure of winning or being placed.

Some show organisers have a tendency to put bouncy castles or musical performances next door to the showjumping arena, and for these challenges a well-habituated horse is definitely an advantage! But how do you get your horse accustomed to such things, which aren't part of any horse's day-to-day experience? This is where the concept of a well-adjusted horse comes to the fore; if you have put time and effort into meeting as many of your horse's species-specific needs as possible, so that his or her day-to-day life matches as closely as possible the activities that a wild or free-roaming horse would engage in, they will have built up the mental capacity to cope so much better with any strange object that comes their way. Look at the example below of the desert horses whose managed ethogram closely reflects the natural ethogram, so whilst they don't encounter sand boarders or drones every day, they are not fazed by any such experiences because their hierarchy of needs is well met, and they take these things in their stride; they have a good look (the same as their riders do) and then carry on with what they happened to be doing.

The behavioural concept of context specificity is also crucial when we are considering how to set the best possible scene to encourage horses to jump well and happily. Many riders, including Denise, Tom's owner, have been caught out by training a great jumping performance at home, and then arriving at a show and finding that their horse won't go over the first fence, or won't even enter the arena. This happens because the horse has learned to perform the behaviour of jumping in one context only, and the rider has not taken care to make sure the behaviour becomes generalised. The way to avoid this is to practise set-ups at home: make your fences look as much like competition fences as possible; for showjumping, include uprights, parallels, oxers, gates and walls of different colour combinations, for working hunter include rustic poles, brush fences and stiles. Ask a group of friends to visit and act like a crowd of spectators; place unusual objects around the yard as you ride through it to the arena, and include them in the arena too, such as balloons, umbrellas, large cuddly toys, anything your horse isn't used to. Get them used to these things one at a time at first, by approaching to the point where your horse looks at the object but remains calm; praise them with treats or scratches and move away again. As you repeat this over time your horse will be able to get closer to these unusual objects without batting an eyelid. Now you can start to combine different stimuli, for example the group of people plus balloons, or balloons and cuddly toys. With the jump fences themselves, start with just one and build up to several, one at a time. Mix the stimuli around and start to combine three or four of these items, until you can have them all present at the same time and your horse will still jump a course of fences fluidly and calmly. Take your horse to a few shows where you just walk around the showground, letting him take in the sights and sounds without the pressure of performance. Habituating your horse gradually in this way is the best behavioural approach to safeguarding your horse against unwanted reactions at competitions and gives you both the best opportunity to give your top performance in the competition.

Cross-country riding presents some similar and some different challenges from a behavioural perspective. You still need to habituate your horse to the warm-up area, though it might be bigger than in showjumping or dressage, and crowds might present a distraction, but rather than being evenly spaced around the arena and somewhat distant from the fences, they tend to congregate directly beside a fence, on the outside of the course boundary, followed by a period of time where there is no crowd until the next fence appears. So, recruit the same rent-a-crowd of friends, but have them safely stationed closer to the fence, and make your schooling obstacles look more like cross-country fences. As well as stand-alone obstacles, cross-country fences are often set into field boundaries, so set up some of your schooling fences like this, by setting up three or four fences next to each other in a row across the short side of the school and riding each of them in different sequences. Remember to build up the height and spread of your schooling fences gradually, in stages where your horse can succeed at each level.

Another relevant behavioural factor in cross-country is that you are riding much further away from the other horses, and you are riding at speed. In the horse's natural ethogram, if they were in this situation it would be because they were running to escape some kind of threat; this can create a fear response in our performance horse even though we know there is nothing they need to escape from; merely the fact that they are in the same set of combined stimuli (alone and galloping) can trigger this response from a neurobiological perspective. This is another problem that Denise encountered at a cross-country competition with Tom, where he refused to leave the other horses, even before he arrived anywhere near the first fence. In behavioural terms, it's always better to prevent problems happening, rather than to try and correct them once they have arisen. Using the concept of shaping, we can practise this at home by increasing intensity and duration to prevent this herd-bound type of problem occurring, and in this example we need to be aware of context, speed and distance: first, ask your horse to leave one other horse and move away a short distance. Gradually and over time, increase the distance. Then, increase the number of horses one by one but reduce the distance again and build it up. Reduce both these stimuli but change the location (context) where you practise this set-up; build the other two stimuli (number of horses and distance) back up again. Reduce all three stimuli and start to move away at increasing speed: walk, trot then canter. Build this "compound stimulus" (a group of stimuli) back up until your horse is happily cantering away quite some distance from a group of horses. The more that you have practised this in different contexts, the more relaxed your horse will be about leaving other horses to set off on a cross-country course, so practise this when you are out hacking with your friends, especially if you find a small ditch or log that you can pop over.

It's always a good idea to have a go at any easy obstacles you come across when you are out riding in company; you and your horse will both be in a more relaxed frame of mind and in these conditions new tasks can be easily learned and transferred to long-term and procedural memory. Denise started doing this when she felt confident about riding Tom again, and it paid dividends when she rode in competitions later. You can also hire cross-country courses to practise over, and again take opportunities to ride *hors concours* at competitions. Pairs cross-country is a popular way to introduce young or novice horses to this activity too, where the newcomer can follow a calm and more experienced horse around the course. Denise's friend gave her some very good advice when she suggested that she enter Tom in a pairs cross-country competition after their first attempt did not go according to plan.

Hunting and horse racing both give domestic horses an unequalled opportunity to gallop very fast together as a group over a long distance. Whilst this behaviour is part of their natural ethogram, the difference here is that the group is not a family or bachelor group, or a familiar herd. We can look at the types of occasions in nature where horses would gallop together: escape from threat, or play, each situation accompanied by a very different emotional state. Where several horses are attempting to escape from a significant threat, they may join together with other bands they don't know very well but probably have some knowledge of, given their ability to recognise sounds and smells over vast distances. An example of this is free-roaming mustangs in the United States running away from a helicopter round-up. In this situation, the horses will be experiencing a very negative fear emotion (Panksepp, 1998). However, horses within a familiar, bonded group will gallop together for a very different reason with a different, positive emotional connotation, when they are engaging in one of their innate play patterns known as "chase and charge". Here, they will suddenly run from one point to another, stop and look around in a very alert and excited way, then all start running again, stop again on full alert, and so the game continues, until it finishes as quickly as it started and the group members all put their heads down to graze calmly again. Considering hunting and racing, neither activity completely reflects the "chase and charge" game although each has some aspects of it. It could be argued that in racing, the horses are galloping to escape the threat of the whip; however, the whip is typically only used in the final stages of a race, and racehorse trainers know that young, fit thoroughbreds don't need any encouragement to run fast together, whether under saddle or in a field. From a behavioural perspective, it is likely that the ethological effect of behavioural rebound is a contributory factor here, where a behaviour that is typically performed at a certain intensity, frequency and duration is inhibited; in the case of most racehorses, fast movement, which they are motivated to do, is inhibited by being stabled for parts of each day, so when they come out on ridden exercise or to race, the rebound effect is what causes them to want to run. Many racehorse trainers approach their training in a behaviourally-minded way, understanding the need for regular turnout in groups, and their horses benefit from spending all or part of the day outside, grazing, loafing and playing with their stable mates.

Habituation to starting gates and tapes, noisy cheering crowds, transport, unfamiliar stabling and racetrack rails will help racehorses to cope better with their experience at the racetrack. Their first few trips out can just be to experience the different atmosphere, before they run. Somebody who takes a behavioural approach to racing is jockey Megan Ellingworth. Megan rides Barnsdale, a five-year-old thoroughbred who currently runs on the all-weather circuits in England. Having researched the benefits of a barefoot approach, Megan took Barnsdale's shoes off during his summer holiday and noticed that his feet became stronger and he tripped less. Megan says:

His feet now work as nature intended and they naturally absorb shock. So far, I haven't identified any detrimental reason to revert back to metal shoes or racing plates. I coupled this transition with a low-sugar and starch diet, which still provides the required levels of energy, and extensive turnout. The aim is to balance performance with healthy longevity. Initial results are very similar to the previous year but with no abscesses, pulled shoes, nor lameness, all of which occurred previously. I am very much the exception in the racing world and am hoping by being different I will make a difference to the overall health of a beautiful, powerful creature in my care and protection.

Horses who go hunting will benefit from set-ups at home where a number of riders canter across country together (this benefits horses who do team chasing, too) and jump different obstacles. A very useful exercise to practise at home is waiting at fences, as there are often queues to jump fences one at a time out in the hunting field, and gate opening is a helpful skill to learn too. When you take your horse hunting the first few times, go with an older, experienced horse, attach a green ribbon to his or her tail (signifying an inexperienced, "green" horse) and only stay out a short time at first; go home before you feel that your horse is overfaced, or tired, or becoming anxious. Mock hunts and some packs offer hunting lines where jumping any fences is optional, so take this opportunity the first few times as part of the behavioural shaping process.

Hacking and endurance

Most riders want to hack out on their horses at some time, even if their main focus is one of the competitive disciplines. Hacking serves

Photo: Jenni Nellist.

the same purpose as other ridden activities by providing opportunities for physical conditioning, cognitive learning and positive emotional experiences. Horses benefit behaviourally from the enrichment that a hack out provides, especially when accompanied by friends from their turnout group, and it allows them to develop a wider home range and explore it in a relaxed and safe way. It's a great opportunity to travel across a distance that is closer to the proportions of a home range; even a 10-acre field doesn't represent a full home range, and whilst out hacking horses can explore, browse and graze in areas that they wouldn't otherwise have access to.

When we think about the things we see and the places we go when we ride out it is clear that there are a lot of different sights and sounds that it would be helpful for our horses to habituate to, including different road vehicles, different surfaces underfoot, unfamiliar animal species in fields and farm buildings, especially pigs, walkers, dogs, cyclists, trail bike riders … we can all add our own examples. If we have a young horse, or a horse that's new to our premises, it's a big expectation to ask them to work out the riding signals that we are trying to give them, at the same time as taking in so many new stimuli on the ride. We can help them to adapt to this by taking them out on short walks in hand first, to allow them to explore these new places whilst remaining close to their group members for safety and security. As they become more confident, we can increase the distance and variety of the in-hand outings.

It's really helpful to understand a horse's startle response when we're out riding, typically referred to as shying or spooking. This is an innate, adaptive response that has evolved in this prey species, whereby they pay attention to anything that they perceive as a potential threat, they stop to look at it, evaluate and make an almost split-second judgement (we would say, in human terms) as to whether it is safe or whether they need to run away to create distance from the threat (Budiansky, 1997). When you're riding your horse, you'll see his head go up, feel him tense up and feel his forwards movement decrease or momentarily stop. If you're leading your horse, you'll see his head and neck go up and he'll appear to grow a couple of inches. If his head isn't too restricted, you'll see him raise and lower his head as he attempts to take in all the information he can with different parts of his eye, which processes visual stimuli differently. Horses will orientate towards the perceived danger to take in as much sensory information as possible, with their ears stiffly pricked forwards. If they've decided the Thing (plastic bag, bird in hedgerow, tractor, dog, etc.) is safe, they will carry on, often arching their neck so that they can focus the right part of their eye on the object to keep looking at it as they walk past. If they don't think it's safe, they will stop and may spin round to escape in the opposite direction, or they might jump sideways to create more distance, the behaviour that panics car drivers when horses spook at something by the roadside.

Some of the advice you will receive about dealing with these situations includes:

- turn the horse's head away from the scary object so they can't see it;
- keep your outside leg on to make him walk past it;
- never dismount to lead your horse past something they are frightened of;
- kick him on until he goes right up to the object;
- drop your reins and let him sniff it;
- dismount and lead him past;
- find an alternative route to avoid the scary place completely.

Some of these points are more helpful than others. The first one won't help your horse learn anything about the object, good or bad, if he can't see it. The second risks flooding him and making him no less fearful on the next encounter, as does the advice to kick him on to try to get him to move closer. It's fine to dismount and walk past with your horse, this is likely to give him more confidence and can be part of a shaping process to desensitise him to the thing he is scared of, which can also include letting him sniff it in order to take in more information. And if your horse's reaction risks the safety of either of you, the best plan for the time being might be to take a different route until you can work on a desensitisation programme.

Something else your horse will often do out hacking, either in response to a fear-inducing stimulus or because it's simply toilet time, is defecate. Many people think we need to drive our horse on at the pace they're going at whilst they eliminate, but this isn't at all necessary and it's better to let them stop if they want to, as they would naturally do. If we consider defecation from a behavioural viewpoint we learn that when and where horses eliminate forms part of the way they communicate with each other by leaving important deposits of information.

Separation anxiety can be an issue that many would-be hackers contend with, and when anxious horses nap and even rear or bolt this is a dangerous situation. As a group-living species it's not surprising that they become upset when they have to leave other horses; horses have evolved over millions of years to form close relationships with other herd members in order to keep the whole herd safe and secure so that individual members can survive and reproduce. A horse who finds itself alone has no safety mechanism to rely on and is therefore constantly on the alert for any threat, hypervigilant on rides and unwilling to leave the safety of the yard. Separation anxiety can be difficult to resolve but it can be achieved in a behaviourally-minded way by working with the horse's natural instincts and the principles of psychological learning theory, using a retraining programme modelled on what behaviourists term desensitisation and counter-conditioning, which means setting up situations that reduce the intensity of the anxiety reaction and change the underlying emotion of fear so that the horse only has good experiences and is able to remain calm (Bowen and Heath, 2005). This is achieved by making the time spent apart from other horses very short at first and gradually increasing it in small increments that they can cope with without becoming anxious. The horse learns to predict that the separation will never be permanent and they will always be reunited with other horses eventually. The programme of relearning has to take place incrementally at a pace dictated by observable changes in the horse's behaviour; it can't be rushed or forced because the change that is taking place is happening at a neurobiological level, strengthening new synapses and weakening existing ones.

Endurance riding, whilst requiring a greater level of physical fitness than hacking, is often viewed as behaviourally similar, but there are different issues to consider: serious competitors often travel long distances to participate, and this will influence the horse's mental and physical condition as we have seen above. Even travelling to a nearby pleasure ride will have an effect; in both situations, the horse is in an unfamiliar area rather than his home range and this will increase anxiety and vigilance, compounded by often being the only one from his group who has made the journey, therefore he doesn't have the security of his friends to support him.

We often hear the term "happy hacker" used, as if "just hacking" is a lesser activity than other disciplines; however, to hack out safely and happily in a way that is enjoyable for both horse and rider requires just as much training as that required in other disciplines, if not more, because the range of objects that the pair might meet, or potential unusual events, is greater. Often the rider has less control over these encounters; let's take as an example a young horse bought to showjump and compare it with a horse whose owner wants to ride it out across a 40-mile radius including farmland, hills, woods, roads, tracks and suburban environments. The showjumper might be schooled in his or her home arena and travel by horsebox to clinics and competitions where their rider starts their showground education by riding them quietly around but not competing, and building up slowly through smaller classes until they are ready for bigger courses. In contrast, the horse who goes hacking is expected to accept a far wider range of unexpected events, from combine harvesters, tractors and motorbikes to loose dogs, lawnmowers and llamas.

If we think about hacking in this behavioural way, we should be very proud of our amazing hacking horses!

Driving

Driving encompasses a wide range of sporting, pleasure and working activities including carriage driving, work in forestry and agriculture, and various forms of harness racing and competitive driving ranging from lightweight sulky or cart racing, through indoor or outdoor obstacle courses to contests of strength where horses pull heavy weights over short distances. In many parts of the world, people still use horse-drawn transport as forms of haulage and transportation, and in some countries, horses are used to pull ceremonial and tourist carriages. Horses in harness make appearances at horse shows too, whether it's the Shires or Clydesdales used in demonstrations by English breweries, for example, or the finely bred show horses driven in competitions such as the Concours d'Elegance.

From a behavioural perspective, the horses used in all driving activities should be habituated to the vehicle they are expected to pull, and the different pieces of harness that are used for driving which don't appear as riding equipment, including the neck or breast collar, swingletree, traces, vehicle shafts and breeching straps; the latter can be particularly worrying for horses as they go around their hind legs, which is a high risk area for this prey animal. In terms of training, we need to know that functionally the draught horse pushes into the collar, it doesn't "pull" the carriage behind him. There are some significant technical ratios involved in identifying the "balanced draught" lines on each individual horse so that he can draw the weight of the load through his own centre of gravity and it's important to be aware of these so that the horse has a comfortable driving experience as this will minimise the risk of any behavioural problems presenting (Lee, 2004). We also need to consider the effect of blinkers on the horse's vision; horses rely on their nearly 360° vision as their primary sense in early detection of threat and to deny them this facility puts them at a distinct disadvantage which they need to adapt to, both behaviourally and biologically. They therefore need to be given ample time to become familiar with this unusual feeling.

Long reining, or long lining, can be included in the driving category; this activity is becoming more popular and is good behavioural practice as a stage in the training of riding horses so that they can learn school movements and vocal cues before the trainer transfers these to the ridden stage. Here, factors such as the position of the trainer, and the feel of the long lines snaking above the hocks, are elements that the horse will need to get used to. Long reining is an ideal way to exercise young or rehabilitating horses.

The different types of competitive driving present a variety of behavioural challenges to horses, not least being permanently chased by a noisy draught vehicle! This is also something that riding horses can often react to, seeing another horse being pursued in this way. Slow and considerate training is essential so that the horse stays within their comfort threshold in order to habituate to equipment and learn the cues and tasks that are required. Stadium harness racers should be helped to get used to items such as the over-check rein and the different starting gates or tapes, and pacers (where the legs move laterally rather than the usual diagonal pairing) often wear hobbles to support their gait; they can be habituated to these under saddle during exercise rather than limiting their use to the racetrack which would be more likely to create anxiety and worry. Stadium harness racers often spend several days at a track before a race and, with behaviour in mind, this is helpful because it helps them to recover from the long journey and they can become familiar with their new surroundings before they are asked to give their best performance. Training sessions on the track before the race can be used to get the horses used to the moving starting gate which is towed round the track for the first circuit and then departs. Something all racehorses need help habituating to is the noise of crowds; this can be done at home by playing recordings of crowd noise, and for an even better behavioural effect, these recordings can be played at the racetracks during quiet times, whilst being paired with positive experiences such as food, scratches or the company of a friend.

Polo

Polo is an intense game and the ponies involved possess power, balance, skill, coordination and quick reflexes. Enthusiasts say it takes years to learn, decades to master and a lifetime to perfect. Eight players on horseback try to hit a small moving ball with a four-feet long mallet from the back of a galloping horse. At the same time that one player attempts this, their opponent is trying to prevent them from doing so.

A period of play within a polo match is called a chukka and lasts seven minutes and 30 seconds. At the end of each chukka, each player changes to a fresh horse. Players use multiple horses and at the higher levels, players switch horses every three to four minutes, maybe more, to keep the advantage of a fresh horse. A typical six-chukka grass polo match on a standard 300 x 160-yard field will involve at least 50 different horses.

Polo ponies exceed 14.2 hands but are traditionally referred to as ponies. They are specially trained, highly intelligent and tolerant and once they have finished their playing career many of them excel in other disciplines because they have been accustomed to strange environments and situations during their adventurous lives, habituating to noise, yelling and bumping and cheering.

Polo ponies need to learn specific skills for their game, such as accommodating their rider's mallet and ball play, and galloping short distances and stopping and turning at great speed. Riders use a considerable amount of tack to control a fresh pony for the short duration of a chukka. A full contact sport, polo has strict safety rules governing how and when contact may occur and the number of accidents is low in comparison to racing and eventing.

Polo ponies have the behavioural advantage of a very short working period of just over seven minutes, after which they return to their familiar "string", as a group of polo ponies is called. When they are out on the field, their friends are never far away and they are usually exercised in a group, often three or five at a time using "ride and lead". At home, they are typically turned out together as one or more groups, and in the UK they get long winter holidays outside the May to September polo season when they are roughed off and turned away to enjoy simply being horses, although "arena polo" is now becoming popular to extend the season into winter. It is played with a softer ball on a smaller pitch, and this may begin to have a behavioural effect if the ponies don't get the winter rest time from which they currently benefit.

Behaviour in mind, Bedouin style

Bedouin brothers Saleem and Salem Alzalabieh are renowned for their exhilarating trail rides through Wadi Rum, a UNESCO world heritage site famous for its stunning sandstone and granite massifs, towering sand dunes, wide, sandy valleys and narrow, shady canyons. Their grandfather, Hadji Attayak, rode with Prince Faisal and Lawrence of Arabia, and lived to be over 100 years old. He was one of the first Bedouin to join the Arab Legion and was the chief of King Talal's bodyguards.

Saleem and Salem were born in the desert to a father who loved all the distant places: preferring solitude, you might find his Bedouin tent far to the south in Wadi Saabit, or perhaps in the north near the Barragh Canyon. So, his sons know the hidden corners of the vast Wadi Rum desert better than many others of the Zalabiah tribe. Men from their family have always spent much of their time on the mountain summits looking for forage for camels or taking goats higher up the mountains in the summer when there is no pasturage below.

From the beginning, the brothers were taught to survive in the desert. They knew where to find snakes and scorpions, how to stay away from them and what to do if anybody was bitten. They learned how to take care of the tribe's animals and how to ride camels and horses, and they learned all the different plants in the desert and how to use them for their curative proprieties. They know in detail all the wild animals and from their tracks they can recognise the species, its size, speed and how long ago the track was laid down. They can even recognise a particular horse or camel from its tracks. Saleem and Salem's deep knowledge and love of their desert extends to the desert horse, the Arabian.

The brothers' 18 Arabian horses live outside in the desert all year long. By their Arabian breeding, all the horses are spirited, clever and sociable. They are sensible and sure-footed, but like all horses with Arabian roots, they are fast and sensitive and require tactful riding. Each horse is different and living in the security of a settled group, each is allowed to develop and express their own character. Some are Bedouin Arabs, others are pure Arabs and

Anglo Arabs, ideal for the climate and terrain. Their physical environment and psychological well-being keeps them in peak condition for the long trail rides of six hours a day over six days. The desert ground keeps their feet short and hard, so they are unshod and sound. Living as a settled social group, the herd enjoys a high standard of welfare which enables them to cope with the desert climate: they can interact freely at will, including playing together and mutual grooming. They also have free access to resources such as water, shade, shelter and sand baths. Their good physical and psychological condition makes these horses excellent endurance mounts, strong to travel long distances and responsive to ride on lightweight aids or voice cues.

Saleem and Salem are now breeding the next generation of strong, intelligent, sociable young Arabians who learn the ways of being a horse in the herd before graduating to trail riding work. The brothers know from many years of experience what a horse needs to excel on the trail, and they make sure that all the conditions are in place to allow this excellence to show through in their horses in a behaviourally-minded way. The horses live as a family group, including two stallions, mares and youngsters ranging from foals at foot to 5-year-old adults; foals or yearlings are weaned naturally just as in free roaming groups, when the next newborn comes along; this group make-up compares well with the natural ethogram of this species and the security that this presents gives each horse a resilience to be able to cope with whatever comes its way at home or on the trail. This makes these horses very safe to ride; they have a naturally developed startle response which is innate to the behavioural repertoire of horses; it typically consists of stopping, turning the head and orienting the body towards the object that startled them, but rather than escalating to any more extreme reaction, the horses simply carry on with whatever they were doing, rather than expressing a fear reaction such as rearing, bucking or bolting, because their living conditions always give them an underlying confidence and security.

The strong overlap between the family structure of these trail horses and a free-living group also makes the youngsters extremely trainable, primarily by habituation and classical association. Training starts when a foal is a few months old when a groom will lead Mum out from the paddock a few 100 metres into the desert, and the foal naturally follows at foot. As the distance and duration increase and the directions the walks take change, the foal gradually habituates at a manageable rate to its nearby surroundings. This process continues throughout the first two years, travelling further and further from the larger group and gradually transitioning to the youngster accompanying the groom on its own, without its dam. Carrying out this process in this way also enables the group members remaining in the paddock to habituate to the departure of one of the group members and this augurs well for the time when some of the group will leave for several days to go out on trail rides. Through classical conditioning, the foal acquires the same voice cues that its mother responds to, and it learns to stop, go and turn at the groom's vocal request. At the age of around 2 years, the young horse will start to accompany its mother and some of the other horses as the older horses are ridden out into the desert, initially for an hour and gradually for longer periods into different and unfamiliar places; this continues the process of habituation whilst the horse still experiences the secure presence of some of the other horses who accompany it. Eventually, at 4, the young horse is backed, in the company of the other group members and in familiar surroundings – an easy and uneventful process because of the strength of learning that has already taken place and because of the familiarity and positive associations of the environment in which the event takes place. The young horse already knows the voice cues for go, slow down, stop and turn, and as a trail horse this is all they need to know. Now their ridden training begins; following the same process as when they followed Mum out into the desert as a foal, they accompany another horse a short way out and this distance gradually increases along with the horse's confidence and experience. Although they can be ridden by voice commands and weight aids, they are now also ridden with light bit and leg pressure, so that they form associations between the voice/weight cues and leg and bit pressure. In this way, they learn the aids that will be used by the trail-riding guests who typically ride "English" style. Eventually, a groom rides them out with one of the trail-riding groups for half a day, then one, two or three days, and at 5, the behavioural approach taken to their training since foalhood means that they are then ready to join the "working group" and be ridden by the guests who come to enjoy the desert trail rides.

Tom and Denise get behavioural help

We left Tom and Denise in an unhappy state where Denise was too frightened to ride him, but knew that because he was stabled he needed to be ridden to get exercise. Denise phoned the owner of the horse she used to ride, who gave her some very important and useful behaviourally-minded advice. She told Denise to move Tom to somewhere where he could live out 24/7, and she advised Denise to only ride out with another horse, ideally an older and calmer horse than Tom. Denise asked around and was put in touch with a farmer who had a 20-acre field that he rented out to seven other private horse owners. The added advantage of this field was that it was part of a former RAF camp and included a gate placed on tarmac hardstanding, it was well drained, well sheltered by hedges and as well as having extensive grazing, it contained trees and bushes for browsing, so environmentally it was much more suited to Tom's equine behavioural needs than confinement in a small stable. Tom settled quickly into his new herd, and Denise saw him change from a horse who threw his head around and tried to get away from her to one who nickered when she called him across the field and came over to her for scratches and affection. After a few weeks settling in, Denise arranged to ride Tom out with one of the other horses in the field and its owner. On her friend's advice they just went for a short ride and over the next few weeks and months these increased in distance and Tom hacked out with other different horses, some from his own group and some owned by Denise's other friends. The rides included times when Denise would ride Tom away from the other horses, or ride down a different track or road and meet the gang at the other end. After a while, she started taking Tom for short rides on his own. She didn't always get it right and sometimes Tom would nap or plant because she had asked him to go too far, too soon. But most of the time she got it right and this approach paid off because nine months later, Denise found herself enjoying hacks out and about through the woods and across parkland, just as she had imagined herself doing when she first fell in love with Tom.

Another problem presented itself for Denise however, and this was the jumping that she wanted to do on Tom. Unfortunately, the jumping skills she was told he possessed when she bought him turned out not to be what she expected, and she realised he had not really learned to jump. She set about constructing a set of jumps and, starting with a ground pole and building up through single cross poles, uprights and spread fences to a small course, she taught Tom to jump at home. Denise was very happy when they set off for her first local cross-country competition, which she had previously won on her friend's horse, and she looked forward to riding the same course on her own Tom. Tom was fine warming up over a small rustic practice pole and in the collecting area near the start box, Denise and Tom entered the box and the starter counted them down, "five, four, three, two, one … you're away, good luck!" and they cantered to the first fence, a small but wide log on the ground. Where Tom refused. Not once, but three times. Denise and Tom were eliminated before they had even jumped a single fence, and Denise had no chance to enjoy the cross-country round that she had excitedly anticipated.

A helpful onlooker suggested they pop over the practice fence again just so that Tom didn't leave the event with a refusal as the last thing in his mind, and Tom jumped it nicely. What had happened to make Tom refuse the first cross-country fence? We can analyse this from a behavioural viewpoint and look at what Tom had already learned to do, or not learned to do, and in what contexts. The only jumps he had training experience of were rustic fences, and this training was always done at home. Whilst animals can generalise to a degree from one situation to another, there has to be enough similarity between the two settings. In this example, the fences that Tom had learned about in the home training sessions were not enough like the first cross-country fence, a log, for his learning to transfer to this new context, for his brain to make the connection. And the wider context was so different; an unfamiliar field, outside his home range, without his field mates and with a lot of unfamiliar horses. Tom literally didn't know what to do, and the anxiety of being in a strange situation played a part in this. Where was he and what was his rider asking him to do? These were the anthropomorphic thoughts going round his head. He didn't know the other horses, but he would still prefer to stay near them rather than canter away from them across an unknown field, because a group of horses represented more safety than being somewhere unfamiliar and alone. But Denise had now learned enough about her horse's behaviour to know that these were the mistakes that she had made, and that they could be rectified through desensitisation and a carefully structured shaping programme to retrain Tom. Denise set about this, finding opportunities to practise jumping small obstacles in different settings, doing some pairs cross-country competitions and building up to clear round competitions and trips out to a nearby cross-country course that she could hire. Understanding and meeting Tom's basic species-specific needs and applying her knowledge of equine behaviour and how horses learn helped Denise to change Tom from an anxious, fearful and confused horse into a happy, confident and trusting riding horse.

Final thoughts

This chapter has provided examples of how a range of equestrian disciplines can be enjoyed in a behaviourally-minded way. The different activities we have explored and the range of situations a horse experiences whilst engaging in them can present challenges or enrichment to its domestic life. A greater understanding of horses' innate behavioural responses helps us to work with them in such a way that they can learn about their environment and enjoy the activities we ask them to do.

References

Baldwin. J.D. and Baldwin, J.I. (2001) *Behavior Principles in Everyday Life*. Prentice Hall, New Jersey.

Bertone, J.J. (2015) Sleep and Sleep Disorders in Horses. In Furr, M. and Reed, S. (Eds), *Equine Neurology, Second Edition*. Wiley-Blackwell, New Jersey, USA. pp. 123–129.

Bowen, J. and Heath, S. (2005) *Behaviour Problems in Small Animals: Practical advice for the veterinary team*. Elsevier Health Sciences, Edinburgh, UK.

Budiansky, S. (1997) *The Nature of Horses*. Free Press, Cambridge.

Cregier, S.E. (2009) Best practices: surface transport of the horse. Available at: https://www.academia.edu/6616417/Best_Practices_in_Horse_Transport_Animal_Transportation_Association_Proceedings (Accessed 23 December 2017).

Fraser, A. F. (2010) *The Behaviour and Welfare of the Horse*. CABI, Wallingford, Oxfordshire.

Horse & Hound (2014) Flying horses around the globe: how it works. Available at: http://www.horseandhound.co.uk/features/flying-horses-aeroplane-463634 (Accessed 23 December 2017).

Kiley-Worthington, M. (1997) *Equine Welfare*. JA Allen, London, UK.

Lee, B. (2004) *Understanding Harness*. Carriage Association of America. Lexington, Kentucky.

Marlin, D. (2017) *The science of rugging horses – what to use & when*! Available at: http://davidmarlin.co.uk/portfolio/the-science-of-rugging-horses-what-to-use-when/ (Accessed 23 December 2017).

McGreevy, P. (2013) *Equine Behaviour: A guide for veterinarians and equine scientists*. 2nd edn. Elsevier, London.

McGreevy, P.D. and McLean, A.N. (2007) Roles of learning theory and ethology in equitation. *Journal of Veterinary Behavior: Clinical Applications and Research*, 2(4): 108–118.

Mills, D. and McDonnell, S. (2005) *The Domestic Horse: The origins, development and management of its behaviour*. Cambridge University Press, Cambridge, UK.

Panksepp, J. (1998) *Affective Neuroscience: The foundations of human and animal emotions*. Oxford University Press, New York.

Chapter 5

The older horse

Kelly Taylor-Saunders

With advancements in veterinary medicine, many horses are now living into their thirties and ponies into their forties. This chapter will not only discuss many of the challenges faced by owners of elderly horses but also will consider the difficult decision of when and how to end the life of our elderly companions. Throughout the chapter, we explore how elderly horses can be managed in a behaviourally-minded way.

Life with Tonka

Tonka was my big, bright bay Warmblood gelding and was in my care for 21 years. He spent the last 10 years of his life living out in a field that provided him with lots of natural shelter and the regular companionship of familiar herd members. Since the age of approximately 27 years old, he was rugged heavily in winter and unless it was very sunny and dry, he needed a rain sheet to protect him from any wind and any sudden changes in temperature.

He spent his last four years in part of the main field but separated from the main herd during the day, so that he wasn't pestered by the younger gelding's requests to play – it's not fun to have your hocks nipped by a 4-year-old when you are arthritic, quite blind and a bit deaf too.

Tonka had only his incisor (front) teeth left by the time he was euthanised and because of this, for many of his final years, he was fed his daily nutritional requirement in hard feed and allowed to graze and browse freely. His feed was tipped onto the ground in dry weather, or on top of scattered hay in winter, as it was hard for him to eat food from the inside edges of his buckets, and feeding this way enabled him to perform some natural foraging behaviours. In his final couple of years, I found it was important to leave his red bucket near the food piles, to help him locate the piles of food due to his deteriorating eyesight. Each evening, herd member Libby was led into Tonka's side of the field and remained there overnight to fulfil Tonka's fundamental need to feel safe with equine company.

This description of Tonka's management provides examples of the sorts of things the behaviourally-minded owner will consider with respect to the management of their ageing horse. We will come back to Tonka's later years throughout this chapter; he was a very much-loved horse.

How old is old?

Some horses will start to show signs of ageing as early as 15 and 16 years of age. The average life expectancy of horses in feral populations varies significantly depending on the population and environmental conditions, but might be as long as 20 or 25 years. Jenni Nellist, behaviourist and contributor of this book, is someone who has spent many years studying semi-feral ponies living on the Gower Peninsula in Wales; Jenni notes that sometimes the ponies live as long as 25 to 30 years with little intervention other than yearly worming.

In domestic horses, the average life expectancy is varied because of how diverse their lifestyles can be. That said, horses often reach 30 years of age and ponies often reach 40 years of age. Horses that have been stabled extensively and used in various equestrian activities are likely to have more "wear and tear" on their bodies than those who have led more natural lives. It is also important to consider that as pressure on land availability increases, the costs of keeping horses also increase, and together with the fact that supporting an elderly horse is expensive in time and money, it is becoming increasingly common to euthanise horses arguably "for convenience" at a much younger age than they might reach if additional support was given. This changing culture towards ageing horses is reflected in the insurance industry. Many policies consider a horse to be "veteran" at 15 or 16 years of age, with fewer conditions being covered due to the horse being deemed "aged".

Managing the ageing horse with behaviour in mind requires a detailed understanding of the physical and mental changes that ageing brings, together with innovative ideas of how to meet the resulting change in needs.

What changes with age?
Dentition

In order to make up for wear from chewing, molars continually erupt from the gum, until approximately 5 years of age. From this point onwards, the teeth are continuously worn down for the remaining years of the horse's life. At some point in a horse's life, one or more teeth might break. When the horse reaches their twenties, teeth often end up with smaller reserve crown and roots and subsequently fall out, often leaving bare gum and a crevice where food can get stuck and cause further problems. Hooks develop when the opposite tooth is uneven, damaged or missing and therefore the opposite teeth are no longer able to wear each other down. Hooks can be very sharp and painful, leading to sore cheeks and sore gums. Furthermore, worn teeth do not grind forage as efficiently as younger teeth and this can lead to the horse not gaining as much nutrition from a certain amount of forage as they used to.

To support the ageing horse with respect to dentition, regular dental checks are vital. Some horses require sedation to allow this to be done properly, although behavioural work with the horse to prepare him/her for the dentist's visits is invaluable for better long-term outcomes. Many horses cope well being examined without sedation if the dental practitioner is calm when handling horses and allows them freedom to move, even with the dental gag in their mouth.

Elderly horses often need forage in the form of hay replacements and as the field of nutrition rapidly advances, new products are becoming available. Regularly seeking advice from the vet, feed companies and an equine nutritionist is useful in ensuring that the elderly horse receives the most suitable diet for its changing needs. Older horses often require dampened feed to reduce the risk of choking and to make the food more easily digestible. Elderly horses usually take longer to eat their feeds and often require larger feeds than younger horses. Therefore, providing the older horse with a quiet, safe place to eat their meals will avoid them rushing to finish before other horses move them off their remaining feed and will ensure that they are able to eat their whole ration.

Damien Greenshields CEq/D is qualified in advanced equine and animal dentistry. He recommends that horses' teeth are checked by a dental professional ideally every six months. He explains:

> Six-monthly checks are important for young horses and elderly horses, or those that have been found to have any dental abnormalities during previous visits. It is also important that the horse is seen any time that the owner can see the horse has difficulty eating or notes a change in their behaviour.

To support older horses, he advises "Horses may require softer feeds and soaked hay as they get older. Soaking feeds may also be necessary as is making sure that the elderly horse can eat in a safe and quiet place".

In the case of Tonka, when his teeth were worn he needed large feeds of 15 kg of soaked food per day, which he took a long time to eat. I found that as Tonka aged, the behaviour of the herd changed – most of his herd companions were able to move him off from his bucket and eat his food, which was why he was eventually separated from them for part of the day.

Digestion

As mentioned above, poor dentition can affect the way that food reaches the stomach and moves through the digestive system. Older horses might become prone to colic because their sensitive digestive tracts cannot cope well with changes in feed or even in types of grass at various times of year. We can minimise this by keeping a stable environment for ageing horses. Sluggish gut motility can lead to impactions occurring and tumours can also develop. Chronic parasite damage can also cause colic and might result in loose or sloppy stools. In supporting the digestion of our ageing horse, we should carry out worm egg counts and worm when appropriate, however, not all internal parasites can be diagnosed by a stool sample. It is also important to bear in mind that elderly horses can be sensitive to many substances, including drugs, therefore discussing with your veterinarian about when and what to worm for will help you to know what is needed.

Tonka was particularly prone to bloating and colic when he was allowed onto part of the field that had been previously rested, unless his time there was restricted initially; the vet described this as the feeling we get after "Christmas dinner". Therefore, I was very careful not to let him have more than an hour or two on new grass.

Shape

As horses age, their shape and confirmation change. An early sign of ageing is the development of a "sway" or dipped back, when the saddle area sinks, making the withers become more pronounced. This occurs due to weakening of the supraspinous ligaments that support the back. Such changes can lead to tack and rugs no longer fitting well, which could in turn lead to behavioural signs of pain and discomfort. If ignored, these might develop into more serious behavioural issues surrounding catching, general handling and being ridden.

As horses age, their muscle mass changes and muscles start to atrophy as the horse becomes less active. In some breeds, or in horses used for certain practices, it is more common for the

Photo: Suzanne Rogers.

fetlocks in the hind legs to "drop" due to degradation of the suspensory ligaments. To support ageing horses with respect to their changing physical abilities, it is especially important that the herd dynamics is not such that chasing, play and active behaviours could cause the elderly horse increased risk of injury. Sometimes older horses do well in smaller acreages than younger horses; if in a larger acreage, they might struggle to keep up with the herd, especially if the herd can go out of sight. This can cause the elderly horse to become anxious and become physically exerted as it tries to keep up.

Ageing horses can also find it difficult to lie down and get up again, and to rest, sleep or roll due to general weakness, changes in conformation and arthritis. Speaking with your veterinarian to assess and manage any pain is a vital part of managing the ageing horse. Sleep will undoubtedly be affected if the horse is unable to lie down and elderly horses have been observed to rest and sleep more than younger herd members, as is the case with many species, including humans.

We can support ageing horses by providing an appropriate surface for them to stand and sleep on. In winter, deep mud or frozen, poached ground can be particularly difficult for elderly horses to navigate, causing injury or putting them at greater risk of being hurt.

Making sure that resources are well spaced out also allows elderly horses to eat hay piles in peace, rather than being moved off by younger horses that may be higher than them in the social hierarchy. Situating the water source away from corners of fencing or gateways, will ensure that the ageing horse can drink when they need to, without the threat of being cornered.

Removing and replacing rugs and checking for sore or bald spots daily in winter will allow ageing horses to stay comfortable and allow owners to check for any changes in body condition score. As well as using a weight tape, take regular photos of your horse from various angles and score his condition as this will be beneficial for keeping track of any changes. It is often easy to miss changes in shape and weight when we see our horses daily, but keeping a photo and video library can allow accurate monitoring of the older horse.

I found it easier to keep Tonka a comfortable temperature using layers of rugs instead of one heavy one during winter. This meant that layers could be adjusted and removed according to his needs and the use of removable neck sections allowed him more freedom in movement than full neck rugs.

Arthritis

Some horses that have had an active career can experience "wear and tear" to their joints. Poor confirmation can also have a role as more pressure and strain will be placed on certain parts of the body. Pain can prevent horses from being mobile, including being able to get up and lie down easily. This might affect their ability to perform normal and necessary behaviours such as rolling and lying down both on their sternum and flat out (laterally recumbent), which are important for sleep (Figure 5.1). Experiencing pain can also mean that the horse might be less able to move out of the way of herd members, especially at times of unrest within the herd, such as when a new horse joins or when resources are scarce.

A behaviourally-minded approach to supporting the elderly arthritic horse would be to speak to your veterinarian about assessing and managing your horse's pain and ensure that your farrier or trimmer is able to take into consideration that your horse may now struggle to balance and hold their foot up to be picked out or for trims. It can help to give your horse a break in between feet and attend to a different horse, which will allow the elderly horse to rest. Providing a deep, soft bed to lie on in a stable (if kept in) and allowing plenty of space to roll and get up is also important. Ensuring that the herd

Figure 5.1: Elderly horse Jak showing he is still very capable of the maintenance behaviour of rolling. Photo: Catherine Bell.

is stable and making sure that resources are well spaced out and plentiful will also reduce the potential of any agonistic behaviours between herd members.

Keeping warm and being able to move freely is helpful for old joints and as counterintuitive as it may seem, increased turnout, with appropriate rugging if necessary and access to ad lib forage will allow the ageing horse to keep warm and mobile more than a stable.

Tonka had always been a big mud monster in his younger years and I used how clean he was as a measurement of whether he was able to get down to roll. During the last few years, his rug was much cleaner than it ever had been, but in good, dry ground, he was able to get up and down more easily. During the winter, a deep straw bed was used in the field, to give him a soft place to lie if he chose to.

What to do when a horse is down in the stable or field

One of the most distressing situations an owner can come across when visiting their horse is to find them lying down in the stable or field and realising they are not just resting. In this situation, it is obviously essential to call the veterinarian immediately, who will assess whether lifting the horse is going to be a safe and viable option and will also be able to administer sedation if necessary. Calling the fire brigade might also be necessary. Many teams have been given training in the technical rescue of large animals and will be able to not only assess the situation safely but also use specialist equipment and tools to lift the horse safely.

Sometimes, when a horse is down, it is the result of an accident, but at other times, it is a sign that the horse is too weak to fulfil their basic maintenance behaviours and this is a strong indication that their welfare is significantly compromised. As a prey animal, being unable to stand or move away from threat is particularly frightening and can cause various physical conditions. Horses have not evolved to lie down for longer than 15 to 20 minutes at any one time. Amber Batson, a veterinary surgeon and contributor of this book explains:

> Horses have large muscle groups that are used to move their lower limbs in order to flee. When a horse is down for prolonged periods of time, the blood supply to these large muscle groups can be compromised and this can lead to muscle damage.

Furthermore, she states that when the blood flow does finally return to these muscles, the toxins from the muscle damage can have serious consequences on their kidneys and other organs. Colic is also a risk for any horse that has been down for a considerable amount of time.

Eyesight

In humans, one of the most common signs of ageing is deteriorating sight. It is generally thought that loss of eyesight is not as common in equines as in humans. However, as animals age there will be retinal changes, pigment changes and potentially the development of cataracts due to age degeneration and these might all lead to impaired eyesight, particularly in low light. Given that horses are crepuscular, which means that they are most active at dawn and dusk, it follows that changes in their eyesight could lead to behavioural changes in the ageing horse. Rarely, horses with pituitary pars intermedia dysfunction (PPID) can lose their sight because the tumour on the pituitary gland becomes enlarged and presses on the optic nerve.

In supporting the ageing horse and changes

to their eyesight, we can minimise changes in the environment and changes to where the horse is kept. Having to cope with changing herd members is also very stressful. We can reduce the chances of startling our elderly equines by approaching them from their side, in line with their eye, rather than from behind or directly from the front, so that they have the best chance of being able to see us. Calling to them as we approach can also help, as can making sure that we take our time when putting on head collars or rugs, as sudden movements such as reaching towards their ears or over their backs can be startling.

Hearing

Hearing loss does occur in horses but is often under-reported. Although ageing horses might not experience complete hearing loss, even some degree of hearing impairment can significantly impact them. Horses may be able to hear sounds but not localise where they are coming from and as a prey animal, this can cause anxiety. Horses with hearing loss that has developed over time might have slowly adapted to their environment and their loss in auditory function is not always obvious, unless assessed and diagnosed by a veterinary surgeon. In supporting the ageing equine that might have hearing loss, your veterinarian will be able to perform various tests to diagnose the extent of the loss.

When handling horses with hearing loss, it is often helpful to rely on their other senses such as relying more on visual and tactile cues. As Tonka's owner, I found it useful to gently touch his shoulder or neck as a precursor to putting his head collar on or adjusting any rugs and he was already used to me gently sliding my hand down the back of his cannon bone as a request to lift his feet.

Cognitive dysfunction syndrome (CDS)

This is a degenerative disease that is known to affect most animal species. In humans it is commonly called senility or dementia and there are many sub-categories and diagnoses under the broader terms. Although there are few comprehensive studies into this disease in horses, many parallels can be drawn in the behaviour of some older horses and in other species that have been diagnosed with CDS. As with dogs and cats, some horses stare into space or appear to "get lost", even in familiar areas of their yard or fields. They can generally seem disorientated and confused, which can make them anxious, even about things that they have coped well with earlier in their life. Such changes could also be the result of other conditions so as with all changes in behaviour, it is important to seek veterinary input.

There are specific treatments for CDS in other species but not currently for horses. A behaviourally-minded way to support a horse with suspected CDS would be to keep the environment as familiar as possible, keeping within the same stable and area of the yard, the same herd and same home. Keeping routines predictable can help to reduce any possible causes of anxiety, although one of the signs can be confusion, even in familiar surroundings.

I was mindful of possible cognitive degeneration when making the decision to move Tonka and his herd closer to where I lived. This was not a decision taken lightly and considerable planning was needed to minimise stress for the whole herd. I recall how heartbreaking it was when I walked towards him from the back of the new field one day and he seemed not only surprised to see me but he did not recognise my voice and was startled enough to trot away from me. This improved as he became more settled in the new environment but I always needed to be mindful about how I approached him, making sure that I didn't appear from behind or directly in front of him and called out to him in advance of approaching, to help him be more aware of where I was; possibly a sign of CDS.

Pituitary pars intermedia dysfunction (PPID)

This condition was formerly known as Cushing's disease and is seen mostly in older horses. PPID is caused by changes in a part of the brain called the pituitary gland. Horses suffering from PPID do not produce enough of the hormone and neurotransmitter dopamine, which means that certain hormone levels are unregulated, in particular a hormone called adrenocorticotropic hormone (ACTH). Clinical symptoms of PPID are closely linked to these elevated hormone levels and include laminitis, an abnormal hair coat, which can range from mild changes in shedding or colour through to a curly, overgrown coat. Other signs include abnormal fat distribution (including a pot belly), loss of muscle condition, increased thirst, fat pads around the eyes, excessive sweating and lethargy, and recurrent infections including foot abscesses and sinusitis. Vets take a clinical history, look at clinical signs and might test blood to check several hormones including ACTH levels. (See the website www.talkabout laminitis.co.uk for further information.)

If PPID is suspected, it is vital for owners to liaise with their vet and discuss whether it is time to carry out blood tests to explore the horse's pituitary gland function. Tests might include ACTH levels or "dynamic" tests such as the thyroid stimulating hormone (TSH) stimulation test. You may choose to medicate your horse, but it is helpful to discuss any side effects of medication before starting. If medication is appropriate, your horse will likely be started on medication and once the appropriate dose is reached, the clinical symptoms of PPID usually resolve.

Sleep

As described in the Introduction when we looked at the needs of horses and their behavioural ethogram, horses spend a portion of their day resting and sleeping, some of which must take place lying flat out on their side. To avoid pain from getting up and down, elderly horses can develop slightly different sleep patterns to younger horses, for example they might spend longer lying down at once, rather than multiple bouts of lying down. Sadly, it is not uncommon for elderly horses to suffer from sleep deprivation due to pain preventing them from adopting the postures that sleep and rest require. If the ground is too wet or uneven, this can make it difficult for horses to get up and down easily.

In supporting elderly horses to achieve adequate sleep, your veterinarian should be consulted to assess and manage any pain. Again, a soft bed should be provided, not only in the stable but also in the field. For horses that live out, having a shelter with a deep bed inside, can help them get much needed sleep when the ground is wet. Having enough space to get up and down safely is also important and to be able to do so without any threat from herd members is crucially important. Horses are most vulnerable when lying down and those that can't get up quickly are at greater risk of being hurt. For stable-kept horses, a deep bed will allow the horse to lie down comfortably and the stable must be large enough for them to lie down safely, without knocking themselves as they get up and down.

Various studies have shown that horses spend longer lying down when on a straw bed than other types such as woodchips (e.g. see Pedersen *et al.*, 2004). Straw bedding also increased the occurrence of bedding-related behaviours such as rolling before standing up (Mills et al., 2000). Another set of researchers found that stable size made a significant difference to the length of time horses spent lying down on their sternum (with legs tucked underneath them.) In stables that were 2.5 times the height of the horse's withers, horses were found to lie down for significantly longer periods of time and were more likely to roll before getting up (Raabymagle and Ladewig, 2006).

At around 25 years of age, Tonka was observed falling forwards and onto his knees when he was stood dozing. True narcolepsy is incredibly rare in horses and the veterinarian diagnosed that he was likely falling because pain was preventing him from getting up and down easily and safely. He was sleep-deprived and literally falling into REM sleep whilst standing, which is something that horses must be laterally recumbent for (i.e. lying flat out on their side). He started daily anti-inflammatory medicine and this helped, alongside a course of acupuncture. A deep bed was also provided in the field.

Social status within the herd

Older horses can and should be valuable herd members. They are able to guide younger horses as they learn about the world and how to interact with other herd members. Older horses can be vital playmates and protectors, as well as important teachers to all herd members.

As horses age, they often lose their social status within the herd. This will be particularly evident with free-roaming herds when the stallion will naturally be challenged by younger males, as they bid for alpha status within the herd. Within domestic herds, this is also apparent when the ageing horses are moved away from resources that they have always had easy access to. For example, if an older horse is eating from a hay pile, a younger, fitter horse may use body language and changes in facial expression to move the older horse away from the hay. This might result in older horses having less available to them in terms of resources such as hay or favourite bits of pasture, and in being prevented from being able to stand at the gateway to come in, not being able to drink from the trough and not being able to eat from their own feed bowl. Horses feel safest when they are in a familiar and stable herd. Sudden changes such as those that occur when herd members change, can cause many difficulties, especially for elderly and less able members of the herd.

We can support the ageing horse by ensuring that they have enough space to be able to move away from herd members if needed and that the resources in the environment are plentiful. One large round bale of forage might be enough to feed the herd but if the other horses "guard" that supply from the elderly horse and do not allow them to eat, the elderly horse will not get their share. In these situations, it is important to ensure that there is plenty of hay available and that the hay piles or bales are sufficiently far apart that all members of the herd can eat in peace. If your horse has hard feed to supplement his diet, the safest thing to do is to separate him while he eats.

Wintertime is especially hard and when resources are low; at such times, horses are much more likely to have resource holding contests. This may involve posturing and no contact but in cases when horses are particularly hungry or when the herd is destabilised as when a new horse arrives or suddenly leaves, these contests can involve physical contact; elderly horses might not be able to outrun younger horses and might get hurt. As already mentioned for the other aspects of ageing, older horses can be supported by minimising changes to the herd and the environment.

Separation anxiety

The physical and psychological changes that occur during ageing can impact the horse's ability to feel safe within the herd and yard environment. It may be as simple as the herd moving to a different part of the field, perhaps out of view, leaving the elderly horse behind, that triggers anxiety and stress-based responses in the elderly horse. Separation anxiety can develop because of these changes and indicator behaviours include the elderly horse calling to herd members, stress defecating or trotting up and down the fence line (Figure 5.2 shows a worn path where an elderly horse trotted up and down the fence line when he moved home). Losing their favourite companion can also be particularly hard for elderly horses, especially those who have experienced repeated separations throughout their lives. The attachments that horses develop with particular herd members will usually last a lifetime if allowed to. Sadly, for domestic horses their relationships are often abruptly ended when horses move yards, are sold on or when they die.

Being aware of your horse's favourite companions and trying as best as possible to liaise with their owner so that your horse and theirs can have as much time together as possible, will help your horse to feel secure. Stabling your horse next to their favourite companion, with a window grille between the connecting walls enable horses to see, touch and sniff each other. If your horse's favourite companion dies, ensure that your horse can see the body before it is removed, so that they understand what has happened to them and have a new companion available to prevent them from being totally alone. We will discuss this in further detail below.

Welfare of the elderly horse

Earlier in this book, we have considered the concept of welfare and explored the Five Freedoms, Five Needs and Five Domains models of welfare. It is helpful to keep these welfare frameworks in mind when observing your horse, especially as he ages. It can be difficult to be objective with our own animals – we love them and many of us would probably hope for them to live forever. In order to help owners remain as objective as possible, it can help to complete a quality of life assessment. You can do this alongside your veterinarian if you prefer.

We have also already considered (Chapter 1) the equine ethogram and maintenance behaviours, fundamental equine needs that are essential for optimum physical and emotional health. Table 5.1 considers these in terms of how well they are met in old age. By allocating a score out of 10 for how well the individual horse's

Figure 5.2: A path worn by an elderly horse who had recently moved home and was pacing due to anxiety associated with short-term separation from his herd-mates. Photo: Catherine Bell.

Table 5.1 Considering how well equine needs can be met for elderly horses.

Need	Consideration in old age
Responding to the environment in order to stay safe	Horses need to be able to respond appropriately to threats in their environment. They can only do this if they are pain-free and mobile. Poor eyesight and hearing might also impact their ability to do so safely.
Eating and drinking	Horses need to be able to eat properly, to maintain a healthy condition score. As mentioned above, poor dentition and digestion will impact this.
Rolling and other body care	Horses need to be able to get up and down easily for rolling. Older horses might struggle to roll from one side to the other and might need to get up and get back down on the other side, in order to roll on each side. Horses also need to be able to urinate and defecate easily and without pain.
Rest and sleep	Horses need to achieve appropriate amounts of sleep and there are various ways that this can be impacted by age.
Motion	Although play behaviours reduce in frequency with age, horses still need to be able to move freely and keep up with their herd.
Exploration	Changes in eyesight and mobility might impact your horse's ability to explore his surroundings.
Territorialism	Horses need to be able to maintain their personal space with herd members, which can be particularly difficult with new herd members, when resources are low and if they are not particularly mobile in old age.
Association	Being able to establish and maintain secure attachments with familiar herd members is crucial for emotional and physical well-being.

needs are met in old age, quality of life can start to be explored.

A sad fact is that many horses have a large number of homes in their life; continually being sold on when they no longer meet the requirements of each successive owner. At some point, many horse owners will be faced with the dilemma of only being able to afford one horse at a time and when that horse is no longer rideable, they may want to find a home for that horse, to enable them to pay for another riding horse. If a good home can be found for the ageing horse, this might be an appropriate option. However, there is the concern that passing them on puts them at risk of being sold on again and again, experiencing much disruption and compromised welfare. In their 2012 study, Parker and Yeates noted that when scoring Quality of Life indicators, it is important for owners to know what is normal for their individual horse and that not performing normal behaviours could be an indication that the horse is suffering in some way (Parker and Yeates, 2012). In addition to this, they specified that performing normal behaviours may determine a horse's quality of life, where these types of behaviours are not only enjoyable but also necessary for avoiding suffering.

Behaviourist and contributor of this book, Debbie Busby, described how when her gelding Rusty reached his thirties, he began to show signs of ageing that made her think it would soon "be time," but not quite yet. Bearing in mind the adage "better a week too early than a day too late" she drew up a checklist of maintenance behaviours and began to monitor him several times a day to see which ones he was performing. The behaviours that she wanted to see related to eating, drinking, movement and affiliative behaviour with his friends in his well-enriched field. In particular, Debbie knew that it was essential that he could get down and get up easily, but acknowledged that if she didn't observe this whilst watching him, she could look for mud on his rug, tail and neck as indicators. Debbie soon realised that Rusty was classically conditioned to respond to the sound of her car and would stop what he was doing and walk to the fence line to greet her. This meant that she was unable to observe natural behaviours at such times. So over several weeks she developed "ninja-type" tactics of checking him from a variety of hiding places so that he didn't know she was there and could therefore observe the behaviours he would normally be doing if she wasn't around. Although she says this was an emotional time (as they were together for 27 years), the checklist that she used really helped her to be more objective about his progress, and when the time came to make the decision to euthanise her dear boy, she felt that she was better informed because of it.

Saying goodbye

At any stage in a horse's life, a freak accident, an unexpected bout of colic or sudden laminitis can mean the end has come. However distressing, it is important to plan as much as possible for such eventualities, so that decisions can be made with your wishes and your horse's needs in mind. Who will be responsible for following through with these wishes, should you not be able to be there, for example if you are away or if the horse outlives you, must also be considered.

It is upsetting to think about some potential situations but adding your horse/s to your will is a wise action for any owner to make. Describing your wishes in your will make it clear to your loved ones what should happen once you are no longer around or able to care for them.

The UK-based welfare organisation, World Horse Welfare, have some very helpful documents for this stage of your horse's life. One is called "*Just in case – the facts*" and outlines some of the information that will help owners to make decisions about what they want to happen. The other document is called "*Just in case – plan*", which enables owners to specify their wishes.

For the owner of an ageing horse, some planning is advisable regarding the likely scenario of non-emergency euthanasia. For example, owners might decide that "this" summer will be their horse's last and plan for him to be euthanised in the autumn, before the bad weather comes and he finds it more difficult to cope. A well-known saying is "Better to go a week too early, than a day too late" – summarising the idea that if the end comes when the horse is enjoying life it is preferable to euthanising the horse when his quality of life has decreased to the point it is considered no longer ethical to carry on.

Practical aspects for you to consider when planning euthanasia include making sure that your horse is somewhere where he feels safe, somewhere that is familiar to him, with the herd nearby, so as to minimise any stress. During his last day, you may wish to feed him tasty treats and let him onto long grass for a couple of hours. Many believe that the moment of euthanasia, regardless of method, should be hidden from the herd's view and careful positioning of cars/trucks/trailers can help with this.

It is important to consider the effect that losing a herd member will have on the rest of the herd. It is advisable to allow the herd to approach the deceased horse. This allows them to say "goodbye" and to understand what has happened to their friend, which is an important part of their grieving process. The amount of time that the remaining herd spend with their deceased friend will depend on many factors, including how much time has been allocated between euthanasia and when the body will be collected. A couple of hours seems to be sufficient, although longer can be given to reflect a more natural situation.

When approaching their deceased friend, herd members are likely to demonstrate behaviours that indicate arousal and often conflict. They may be interested to approach the horse but cautious to do so and this may include some approach and retreat behaviours, with high head posture and tail held up high. Herd members may circle the deceased horse, before finally approaching and sniffing the body. Horses have also been observed to lick, chew and even paw at the body on the ground. This is all part of exploratory behaviour that allows them to understand what has happened.

Animals experience grief in much the same way as humans do, as indicated by their behavioural response to loss and separation. Those left behind may experience loss of appetite, poor sleep patterns and seem lost and disorientated. Some may even become clinically depressed and this might also present as physical problems.

We can support the horses that are left behind by spending extra time with them, reducing demands of training or riding, providing enrichment and offering the opportunity for positive experiences to keep them occupied, without tiring them physically or emotionally. It may take several months to years for the herd to settle after the death of a member.

I noticed that Tonka's herd were generally unsettled and easily spooked, for some weeks after Tonka died. In particular, Little Molly, who had known Tonka the longest of all herd members, seemed the quietest and most affected.

Jenni Nellist describes a situation when a dog walker found an elderly mare that she had named Crystal, dead on the common. Jenni was not sure of her age because Crystal had been dumped on the Common several years earlier. Crystal had however foaled for the previous three years and actually had a 9-month-old foal at foot when she died. Prior to her death, Crystal spent most of her time with another aged mare, sharing care of their foals. The last time that Jenni saw Crystal was two days before she died. Her body was found in the core area of the home range and her family appeared to be "carrying on as normal". Crystal's youngest daughter was closely bonded with her yearling sister and had dispersed from the rest of the group at the same time that many ponies were being rounded up and removed from the Common.

After Tonka was euthanised, his herd were allowed to see him. They approached with caution, knowing that something was different as they could see him lying on the grass. The horses approached in a wide arc, at trot and as they got closer, they slowed to a walk and took it in turns to approach one at a time. The mares came over to sniff his body and Bear, the young gelding, began to nuzzle Tonka's body and even pawed at him. Gradually, the herd settled and began grazing on the long grass near to where Tonka lay. Two hours passed before the herd suddenly took off at a trot, back up the field and away from Tonka's body – they went back to eating the hay in the paddock. Soon after this, Tonka's body was collected.

Methods of euthanasia

The method of euthanasia chosen will depend on the individual horse's circumstances and also on what is likely to happen when their body is taken away. A firearm might be more appropriate for needle-shy horses so as to reduce the stress in the last moments of their life. A lethal injection, however, might be required for a head-shy horse. The cost also varies with different methods, with lethal injection being the most expensive option.

Lethal injection

Your veterinarian will administer a lethal dose of anaesthetic straight into a vein in the neck. Prior to this, your horse is likely to have been sedated and they will also have had a catheter inserted to allow the veterinarian easy access to administer the drugs straight into the bloodstream. Once the drug is administered, the horse may respond in various ways: some stagger forwards or sideways, before falling over; some sit down on their hind legs, before rolling to the side; some fall to the side and onto the floor. Once on the ground, the horse's legs and muscles might twitch. Your horse will lose control of their bowels and bladder, a normal reaction in death. The pupils in their eyes will dilate and their eyes will become "glassy"-looking. You will not be able to close their eyes as they do in the movies when a human dies. The changes in their eyes can be upsetting and people may choose to cover the horse's head once they have said goodbye to them. Some horses might also let out what is known as an agonal gasp, which sounds like a breath, even though they will be dead. Your veterinarian will check to make sure the horse is dead by touching the cornea (surface of their eye) and seeing if there is any reaction such as blinking. They will also listen to the heart with their stethoscope to check that the heart has stopped beating.

It is better if the horse is standing on a soft surface when euthanasia happens, so that when they fall down, further damage can be limited. If it is your personal choice that lethal injection is used when the time comes, you may need to train your horse to accept needles and to tolerate being handled by the vet.

Due to increases in health and safety procedures and gun control measures in the UK, lethal injection is becoming the most common procedure for domestic horses.

Free bullet from a pistol

This can be carried out either by a veterinary surgeon or by a skilled and experienced operator. Captive bolt guns are rarely used with horses.

Care must be taken to ensure that the bullet enters the brain at the correct place, in order to cause immediate death. Care should also be taken to ensure the safety of everyone around the horse at that time. If the horse moves his head or steps forwards, the bullet might not hit the correct part of the brain. To support our elderly horse at this time, it can help to give the horse a bucket of their favourite food, so that their head is low and they are distracted. Once the gun has been fired, the horse will usually "snatch" their legs up underneath them in a rapid movement, before falling immediately to the floor. Sometimes their legs might stiffen and they will fall over. Horses die immediately with this method and some owners may choose this for the instantaneous death that occurs.

When this method is used, horses typically bleed from the entry wound in their head and from their nostrils. Once on the ground, their legs might twitch and even "paddle". These are all involuntary movements, even though the animal is already dead. The horse's pupils will dilate and appear "glassy" as they do when the lethal injection is administered. They will also lose control of their bowels and bladder. Many people at this point will prefer to have the head wrapped in a towel.

Once your horse has been euthanised, you may want to ask your veterinarian to remove their shoes, so that you can keep them. You might also want to take a cutting from your horse's mane and tail as a keepsake. There are various companies that will paint your horse's shoes and make horsehair jewellery for you to treasure.

Whether you attend the euthanasia is a very personal choice, with many people choosing to be with their horse in their final moments. Ideally, although despite being incredibly upset, you must still be able to follow any instructions given by the veterinarian, so that you can stay safe and ensure that your veterinarian is able to do their job effectively. If you choose not to be there, then perhaps a trusted friend can take your place, or the yard owner or manager of where your horse is kept. Whoever is with your horse needs to be able to remain as quiet and as calm as possible. Depending on your circumstances, you may want to have someone with you at this time and the British Horse Society (BHS) have an initiative with more than 100 volunteer welfare officers who can be available to discuss the various options with you and be with you at the time of euthanasia.

On Tonka's final day, five weeks before his 33rd birthday, he spent the morning grazing the long grass that had been rested all winter. The herd members were kept in a paddock just the other side of where Tonka was so that they were nearby. When the veterinarian arrived, Tonka was slowly led to a grassy area that was closer to the main entrance gate. He stood quietly as the veterinarian sedated him and inserted the catheter. Once the lethal injection was administered, he wobbled and fell down.

I sat on the ground, sobbing and cradled his large head on my lap. I told him how much I loved him and thanked him for all our happy years together. My husband, father and close friend had spent the morning at the field with me, Tonka and the herd; we talked about all of the special times that we had spent with him and what a wonderful friend he had been to all who knew him.

What happens afterwards?

There are many different options available to owners, but some depend on the method of euthanasia. Those that own their land may like to bury their horse at home. You will need to speak to the Local Authority to obtain permission for doing so, as a survey of local water supplies might be necessary first.

A fairly common option is the fallen stock operator or "knackersman". These are experienced dispatchers who will use a gun to euthanise an animal. They are often used by farmers and will also remove the body. Although an unlikely option, an abbatoir could be considered. The horse will need to be well enough to travel to the abbatoir and the owner might even be paid a fee for the animal, rather than having to pay for the service. In the UK, this option will not be available if the horse has received various medications including typical pain relief, as this means that the horse will not be able to enter the food chain. If the horse's passport has a signed declaration that the horse will not enter the food chain at any point, this will be upheld. Many cremation companies will collect animals from various sites before returning to the cremation site. Please bear this in mind, as it might be an unwanted surprise during an already difficult situation.

Many people liken losing a horse to losing a family member. The grief is real and can be overwhelming. It can help to collect together

Figure 5.3: A lovely photographic memory of Tonka. Photo: Kelly Taylor-Saunders.

some of your favourite photos (see Figure 5.3) and frame them in "celebration" of your friend and the many times that you spent together. Horses can be some of our greatest teachers, and the pain that we feel when they die should never be underestimated. You can also speak to a therapist or your doctor if your feelings of sadness and grief are impacting on your ability to perform daily activities.

Tips for owners by Dr Rebecca Giminez (co-author of Giminez *et al.*, 2008; interviewed for this book).

1. Take a long-term view, taking care of him/her into old age is your responsibility. Don't pass that responsibility to someone else.
2. Take lots of photos when the horse is young and healthy, you may wish to remember them this way.

3. Think about the little things, they will be big things when the time comes.
4. Make sure that copies of your wishes are sent to anyone who may be involved at the time – insurance; vets, including those for second opinions; yard owner/manager; a legal document stating power of attorney is required by some vets in some cases.
5. Have a disaster plan in place in case you are away or can't be reached. Make sure that those who will be dealing with your horse have your permission to end your horse's suffering and fast, if needed.
6. Make sure that anyone who may need to enter the property can do so, gateways need to be big enough for all kinds of vehicles.
7. If the death is sudden, ask the vet to perform a necropsy (autopsy performed on animals), as this will help you to find some answers and also may identify a condition which could affect other herd members.

Final thoughts

Given that many of us will end up caring for an ageing horse at some point in our lives, it is important to understand the challenges that our elderly horses face and the impact that this might have for us both emotionally and financially. Although it is not the most comfortable of topics, preparing ahead and ensuring that those who are close to us and those who share the care of our horses are aware of our explicit wishes, can reduce some stress and anxiety when the time comes to say goodbye to our beloved horses.

Tonka's story shows how elderly horses can be managed with behaviour in mind – it gives some examples of accommodations that can be made to enable elderly horses to lead a fulfilled life right up to the end of their days. Getting old is not a death warrant, but we need to ensure that our ageing horses are as comfortable as possible and are able to perform important horsey behaviours, whilst taking into consideration what changes for them as they age. Growing old is a privilege not afforded to all.

References

Giminez, R., Giminez, T. and May, K.A. (2008) *Technical Large Animal Emergency Rescue*. Blackwell Publishing, New Jersey, USA.

Mills, D.S., Eckley, S. and Cooper, J.J. (2000) Thoroughbred bedding preferences, associated behaviour difference and their implications for equine welfare. *Animal Science*, 70: 95–106.

Parker, R.A. and Yeates, J.W. (2012) Assessment of Quality of Life in equine patients. *Equine Veterinary Journal*, 44: 244–249.

Pedersen, G.R., Sondergaard, E. and Ladewig, J. (2004) The influence of type of bedding on time horses spend recumbent. *Journal of Equine Veterinary Science*, 24: 153–158.

Raabymagle, P. and Ladewig, J. (2006) Lying behavior in horses in relation to box size. *Journal of Equine Veterinary Science* 26: 11–17.

Chapter 6

Riding instruction

Felicity George

Incorporating advice on equine behaviour into riding instruction is certainly nothing new. Around 2,400 years ago, the Athenian Xenophon wrote a treatise on horsemanship (Xenophon, 1999). In it, he gives much advice concerning equine behaviour. He gives many examples of the importance of rewarding desired behaviour, for example "he would receive the bit the more readily if some good should come of it every time he received it" and warns that if a horse is fearful "compulsion and blows inspire only the more fear".

Jumping forwards in time to 1964, Nuno Oliveira's excellent book *"Reflections on Equestrian Art"* (Oliveira, 1988) expresses similar sentiments: "Supposing that a horse is frightened of a certain object and refuses to pass it, turning away violently. Is it punishment that is going to reassure him? I think not. All gentle ways are good, giving confidence and showing that there is nothing to fear" and "reward the horse each time he does what is asked of him. Never ask for more than he is capable of giving".

Throughout the ages, great horsemen have recorded similar advice, recognising areas such as the importance of knowing the emotional state of your horse; helping him to be calm and comfortable; how to gain the horse's attention without force and reward as the most effective form of training.

The importance of the emotional state of any ridden horse is equally recognised today. The FEI dressage rules 2017 (Fédération Equestre Internationale, 2017) state that:

> The object of Dressage is the development of the Horse into a happy Athlete through harmonious education. As a result, it makes the Horse calm, supple, loose and flexible, but also confident, attentive and keen, thus achieving perfect understanding with the Athlete.

So, those desiring competitive success should certainly be training with behaviour in mind, if the above principles are adhered to.

At the grass-roots level, all riding instructors and coaches will have behaviour in mind to some degree when they work with students and horses. For example, every riding instructor who has taught group lessons knows, I think, that relationships between horses influence the order of the ride, and to ignore this courts disaster!

However, some elements of equine behaviour may not be given the priority they deserve. In writing this chapter, I have interviewed four riding instructors who are also equine behaviourists and will draw on their experiences of teaching with behaviour in mind as well as my own. I have also taught equine behaviour to many BHS and UKCC qualified instructors over the years and have noted their feedback on how a more in-depth knowledge of equine behaviour has influenced their approach to riding instruction. In all cases, riders or riding instructors who have then studied equine behaviour have reported that application of a reasonable depth of knowledge of equine behaviour is not a nice extra in riding instruction, but essential in improving both rider safety and equine welfare.

This chapter will:

- explore the benefits of teaching riding with behaviour in mind;
- discuss the elements of equine behaviour that behaviourists who are also instructors have noted as essential to riding instruction;
- discuss how to teach these elements effectively.

As the focus of the chapter is mainly on teaching riders, I will generally use the kind of language I would use when teaching rather than technical terms.

What are the benefits of riding instruction with behaviour in mind?

I regularly taught over 30 riders during a Sunday morning and it was exhausting. So I switched to Friday afternoon, during office hours, as there were only a handful of clients. By the time I left, Friday was booked solid. Client numbers had quadrupled. Meanwhile the Sunday morning trade, under a traditional instructor, had dropped to a handful. I think that's a huge endorsement for a behaviour-focused, educational, riding experience.

(Sharon Smith MSc, Stirling, 2017, personal communication)

Most riders are receptive when they see the difference – light bulb moment … Those who have seen or felt a huge difference are quite moved and fascinated by the change in the horse.

(Marjorie Grant, SEBC PTC, BHSAI, BHS Int SM, UKCC L3, Stirling, 2017, personal communication)

Photo: Anthony Payne.

mments echo my experience and ther equine behaviourists who also teach riding. An improved understanding of how horses learn helps the rider to train their horse with less stress and better results. Helping riders to develop their awareness of how their horse is feeling, and how best to help her feel calm and comfortable when ridden, engages the interest of most riders, young and old, from beginners to experienced riders. Most people get involved with horses because they love them and learning how to work with your horse rather than making her work for you is deeply satisfying.

The two most important benefits of teaching with behaviour in mind are:

- *Welfare:* Keeping the horse calm and comfortable will clearly improve her welfare during riding lessons. It will also impact on her mental and physical well-being outside her ridden work.
- *Safety:* At the last riding school I worked at, I had more than five times fewer reported accidents than other instructors. Vast improvements in safety records were reported by others too; by considering the horse's emotional state and looking after her mental well-being, incidents of conflict behaviours such as rearing, bucking and bolting are hugely reduced.

Another significant benefit is improved client satisfaction. All instructors interviewed commented on this, with those working in riding schools noting significant rises in client retention. Clients who feel safer, providing that they are still learning and interested in their lessons, are much more likely to come back. Most love the feeling of working with a relaxed and contented horse and want more!

These three elements alone should provide ample motivation for both freelance instructors and riding schools to adopt instruction with behaviour in mind. All the riding instructors that I have tutored in equine behaviour now use it as an integral part of their teaching.

What to teach

Almost any facet of equine behaviour may be covered in a riding lesson; as riders become more interested in the subject they will ask an amazing range of questions. In this section, we'll look at the areas that are most commonly covered and are most beneficial for both equine welfare and rider safety.

Throughout this chapter, I will use Emma and Willow as an example of how various elements of equine behaviour can be introduced into riding lessons:

Willow, an 8-year-old Connemara mare, has an issue with the far-left corner of the arena. Her rider, Emma, laughs nervously when she mentions this to me, offering the opinion that Willow just does this to "make me look bad". I ask Emma to talk me through what happens as they approach this corner. Emma describes how she has tried to "make" Willow go into this corner, and Willow's subsequent "evasions". It is clear from Emma's body language and choice of words that she feels this is a confrontational situation with Willow, which Willow started and Emma must win! Emma feels frustrated, anxious, a bit embarrassed and occasionally quite scared. She thinks that she must be firmer with Willow but is worried about Willow's behaviour escalating. I also get some ideas about how Willow may be feeling, but don't want to make assumptions based on just a verbal account.

In this simple and common scenario, many behavioural elements are critical to finding a solution that both Emma and Willow are comfortable with, as we will see in the following sections.

Attitude

When teaching riding with behaviour in mind, we aim to help riders to regard horses as sentient beings, who should always be treated with care and consideration.

Some beliefs about and attitudes towards horses are incompatible with this aim. For example:

- Inappropriate anthropomorphism (e.g. believing horses to be capable of cunning strategies to get the better of us).
- Objectification such as calling a horse "it"; an object that has no feelings.
- Adversarial attitudes. Many riders are given advice such as "show him (the horse) who is boss" or "make him do it or he'll think he's won"; many riders have learned, basically, that they should be having arguments with their horse – and winning them!

This is an area where extreme tact and patience may be needed to help the rider to change. Confronting them head-on is likely to be unproductive; beliefs about the horse's character and motivations, and how to "deal with" her may be deep seated and if new ideas are presented wrongly they may either be rejected outright or cause the rider to lose confidence and feel very insecure.

A powerful tool in changing the rider's perspective is our use of language. In the following passage, Dr Karen Overall is discussing attitudes towards animals with behavioural problems but it applies equally to common attitudes towards ridden horses:

Rather than asking if we can understand behavioral problems as a form of behavioral pain and suffering, or "mental illness", the vast majority of clients and veterinarians, wittingly or not, engage in a terminology and thought process rooted in an adversarial relationship with the animals who share their lives. Physical pain is deemed as "real", afflicting innocent patients; behavioural pain is often thought to be someone's fault or the result of a deeply flawed character ... If what we call something affects the way we think about it, then what we call it is essential; yet we in behaviour have been incredibly careless and in so being, have done harm.
(Overall, 2005)

I wholeheartedly agree with Dr Overall's statement that what we call something affects the way we think about it. Language in common use around horses encourages an adversarial attitude—the horse is being dominant, naughty, evasive, stubborn, and so on. In addition to creating an adversarial environment, many of the terms in common use also imply, as Dr Overall describes, that the horse is at fault or flawed in character.

If, for example, a horse does not respond to the usual aids to bend to the left, but offers an alternative behaviour, she may well be described as evasive. This is seen as "bad" behaviour and sets us in an adversarial situation with the horse. Even worse, it implies that the horse has brought about the conflict by her behaviour. Instead of blaming her for being evasive, we can consider why she gave an unexpected response. Is this movement physically difficult for her; does she understand the aids; is she relaxed enough to be able to pay attention? Then we view this as a problem she needs help with rather than a fault.

Any language that objectifies the horse can

have a profound effect upon the rider. The most common example is calling the horse "it". Shouts of "kick it harder" and "make it canter" and so on are unfortunately heard during some riding lessons. This use of language encourages the opposite of the attitude we wish to foster; considering the horse as a thinking and feeling being. By calling the horse "it" we take the sting out of commands such as "kick it harder". "Kick her harder" might give more pause for thought. Many people will use words simply because they have learned them from others and not really thought about the implications; no harm is meant. But once you become aware of the effect of such language, you would no more call a horse "it" than you would call a person "it"!

If, as the instructor, your language and attitude towards both horse and rider reflects your belief that both are thinking, feeling beings with reasons for their behaviour, the rider will usually be carried along with you, especially when they see the effect this has on their horse.

In our brief introduction to Emma and Willow, we are already seeing Willow being given unrealistic motivations (to make Emma look bad), and descriptions of "evasive" behaviour and "making" Willow behave. This encourages a confrontational attitude and escalating aversive stimuli for Willow as Emma tries to keep herself safe by winning the argument. In Emma's case, she was repeating what she had previously learned, and these were not deeply held beliefs about how horses behave or how they should be trained. Taking the pressure off Emma and discussing how Willow felt and how we could help – as we worked on putting our ideas into practice – was all that was needed to create a quick and fairly dramatic change in how Emma thought about and interacted with Willow. This is not uncommon.

It is interesting to note that the attitude we develop in the rider towards the horse is the same as the attitude the instructor develops towards the rider! We look to help both, rather than labelling and blaming them for "bad behaviour".

Safe places

"Safe places" is really just another term for comfort zones; places where both horse and rider feel calm, confident and comfortable. It is a term that I noticed other instructors using to good effect and have adopted as it is simple and memorable for most riders. Safe places are essential not just for safety but for the horse's welfare; we wouldn't attempt to force frightening or painful experiences on another person and should likewise take care to keep both horse and rider feeling emotionally and physically comfortable.

Safe places depend on many factors:

- That the horse is basically mentally and physically comfortable when being ridden! If, for example, their saddle doesn't fit, or there is some physical issue making carrying a rider painful, then no safe places will be found within ridden work.
- Environmental factors. For example, are both horse and rider comfortable being alone in the arena? With varying weather conditions? With distractions from outside the arena? With all areas of the arena?
- What we ask of the horse (and rider). For example, both horse and rider may feel quite comfortable in walk and this can be genuinely described as a safe place. Trot work may be variable, with both getting tense at times, so not a reliable safe place, and both may be showing clear signs of tension when cantering.

Knowing safe places is extremely helpful to both horse and rider (and instructor!); if we know, for example, that riding a 15-metre figure of eight in a steady trot settles both, then we have somewhere to come back to when tension creeps in, and the rider knows what to do to try and calm things down when she is on her own. This knowledge in itself can be very powerful in calming the rider and building her confidence.

Of course, safe places are unfortunately not static! So many factors influence both the rider and horse and what feels comfortable at one time may not on another occasion. Initially, we aim to give the rider the skills to assess when both she and the horse are comfortable, then to gradually expand the range and size of their safe places so that the rider has a variety of options.

Throughout every riding lesson we are always working on developing the rider's understanding of the horse as a sentient being. Working on safe places is a great way to address this. Initially, the rider seeks safe places because they feel safe to her. As she progresses and sees the parallels in how tension affects her and her horse, she seeks safe places to help the horse as well as herself.

Emma has identified a problem area, in this case part of the arena which Willow does not wish to approach. She doesn't talk about any other worries. However, the first time I see Emma riding Willow, neither look too relaxed! I ask Emma if she'd be happy to try an experiment; I'll lead Willow and keep her in the bottom half of the school, and we can see how that affects her. I suggest that while she is being led, this would be a good opportunity for Emma to focus on her position and movements, and to explore what she can feel from Willow. Within minutes, both horse and rider are much more relaxed, we're finding our first safe place, Emma is improving her understanding of how Willow is feeling, and we can progress from here. Once both are relaxed, I can gradually move away; first, just walking beside them, then moving to a different position in the arena. By the end of the lesson, we have set up cones to mark the area Emma will work within, and I am in the viewing gallery watching them; both are looking calm and confident, and we have our first safe place where Emma can work alone in walk.

Rider foundations

You will always exert high pressures and experience behaviour problems, if you are not a balanced, stable rider. The degree to which you have to be balanced and stable varies from horse to horse. Riding correctly for the horse's physical development will also, by default, result in good mental/behavioural development.

(Sharon Smith, Stirling, 2017, personal communication)

In many riding lessons, an equine behaviourist may be doing the same as a traditional riding instructor; working on the rider's position. When the rider moves well with the horse, the horse will be more comfortable, and horse and rider can communicate quietly with each other. As the rider is helped to develop their awareness of both their physical and mental state, they become able to give clear, consistent and simple aids that their horse can understand. A positive cycle is then established, where both horse and rider become calmer, more confident and communicate with each other in a more civilised manner (less kicking from the rider, less bucking from the horse!).

When the rider is balanced, stable and calm,

they can develop a feel for how the horse is moving, how she is feeling, and the whole experience is much more harmonious.

Conversely, a rider who is not balanced and moving with their horse will be tense, whether she is consciously aware of it or not. The horse will receive multiple aversive stimuli from the rider in the form of, for example, gripping legs, shifts of weight and restraining hands. This will never be conducive to good mental or physical welfare for the horse and may well lead to behavioural problems if the horse becomes increasingly confused and beset by unavoidable aversive stimuli.

> While we are establishing safe places with Emma and Willow, we are assessing Emma's position and movement, and how it is affecting Willow. Emma has identified one corner of the arena as an area Willow won't go into, but in fact Willow is reluctant to go forwards at all. As often happens in these cases, Emma is using frequent and strong leg aids, and is overall, understandably, in a rather defensive position, anticipating trouble. We decide to work on Emma's position, to explore what she could change and how it might help Willow. I take responsibility for asking Willow to move, so that Emma can focus on herself, as in a lunge lesson. Emma relaxes, breathes better, and lets herself move with Willow. At the end of the first lesson, Willow is no longer being inhibited from moving forwards, or confused by constant leg aids, and Emma can work in walk independently. We work on Emma's position without any thought of what Willow is doing, beyond her being safe to ride and relaxed, for the first three lessons.

So, the foundations of teaching with behaviour in mind are fully compatible with traditional riding instruction. I would highly recommend Perry Wood's *"Dressage the Light Way"* (Wood, 2012) and Sally Swift's *"Centred Riding"* (Swift, 1985) for methods of teaching these foundations; achieving a balanced rider who can communicate in a clear, gentle manner with their horse.

Equine body language

Riding, and any interaction with horses, involves communication in both directions. Sometimes there is too much focus on what the rider communicates to the horse without due attention to what the horse is telling us. As you develop a reasonable understanding of equine body language, you wonder how you survived all those years around horses with less awareness!

Every riding instructor I spoke to who is also a behaviourist mentioned this as vital when teaching riding. The benefits are numerous. Safety and welfare benefits have been mentioned at the start of this section and are discussed further below. A good rider adjusts their position, contact and aids depending on feedback from their horse; they are responding to the body language of the horse, and are much safer, kinder and more effective in riding and training the horse by doing so. In particular, recognising and addressing early signs of tension in the horse is extremely important. In all our ridden work, we are looking to establish safe places and gradually expand them, with the attendant benefits discussed above. We need to recognise tension signs to accomplish this.

Recognising mild tension in a horse is a large topic; many riders will recognise tension at the point where the horse is over threshold and reactive but may not notice the early signs. The instructor who is teaching with behaviour in mind will look carefully for these signs;

for example, the slight change in breathing, a momentary tightening of the tail or a change in the shape of the chin as it tightens; and help the rider to recognise these signs. Horses show their discomfort – mental or physical – in different ways depending on many factors; the individual, the type and severity of discomfort and environmental factors. Also, different riders will find it easier or harder to see or feel different signs; for example, some will find it very easy to pick up a slight change in breathing or rhythm, while others find this difficult but will feel a change in the horse's back, or in their jaw. The instructor must really work with the rider here to help them be competent and confident about recognising mild tension in their horse.

In novice riders, achieving this is surprisingly easy. Sharon Smith observes that:

> Even beginners can be taught head carriage, ear movements and tension in the back. I get riders to soften their hips and feel the back moving, then try to allow the horse to move with more activity by softening further. They then want to focus on the horse, rather than just hop on and go fast or jump. The less confident riders become absorbed in the horse, forgetting their own nerves, relaxing and improving their riding without realizing.
> (Sharon Smith, Stirling, 2017, personal communication)

With most experienced riders, this stage also goes very smoothly. Once they are aware of the changes, they recognise them quickly and appreciate the difference in movement in lots of subtle ways.

I would often explain trigger stacking (the effects of an accumulation of stressful events) to the rider at this point, to provide a deeper understanding of the importance of both partners remaining calm and confident. I'd ensure that the rider understands that ignoring the mild signs of tension and "pushing on" can lead to a build-up of tension; something we'd wish to avoid inflicting on another sentient being, and which can also lead to serious conflict behaviours that we'd rather not inflict on ourselves! This is often best explained by getting the rider to think of a situation where they reacted more extremely than usual to a slightly stressful event because they were already feeling stressed. For example, after coming out of a difficult meeting at work, one rider spoke about bursting into tears when someone took the parking space she wanted, even though there were others nearby. Most riders will readily think of some such example, and there is often a "light bulb" moment as they realise how this relates to their horse's behaviour.

Once the rider can recognise early tension signs, they will then want to know how to respond to them! The first thing would generally be to return to a safe place. From there, we address the problem as with any other ridden problem; what are the causes of the tension and how do we help the horse? If the tension is caused by the rider's actions, we can help the rider to change what they are doing. If external factors are involved, we may look at changes to the environment ("we'd really appreciate it if you didn't start up the chainsaw just outside the arena while we're in a lesson!") or look at training the horse, as covered in Chapter 3. The best way to approach this issue will vary widely depending on the rider, the horse, the environment and the causes of the tension.

> With Willow and Emma, we have already helped Willow to be more relaxed by working on Emma; her position, aids and relaxation. Willow is now happy in walk and trot in one half of the arena. We have also dealt with a few incidents along the

Photo: Anthony Payne.

way where Willow has been worried by external factors; for example, a new horse arriving on the yard and being unloaded just outside the arena. Initially, Emma's response to Willow's tension is to become worried too, stiffening her arms and gripping with her legs. Here, I have stepped in to help Emma, and she quickly learns that if she remains calm and non-reactive, Willow settles again quickly.

We now want to start working on expanding the area of the school that Willow is comfortable working in. The area in which she is currently working is clearly marked by cones. In this case, I start by moving the cones about half a metre closer to the "scary corner". I ask Emma to turn right across the arena at the cone and watch Willow very carefully as she approaches the cone. There is no visible alteration in her pace, but her head raises a fraction. As Emma and Willow turn right, Willow's head and neck relaxes again. I ask Emma what she felt, and she noticed the change in head position, and a slight change in Willow's back. We agree to repeat this exercise until Emma feels that Willow is fully relaxed again. Emma correctly identifies when Willow is fully relaxed with this exercise. We then carry on for a further 10–15 minutes in this new space, so that Willow's feelings about working here become more well established. We will then move the cones a small distance again.

How much tension is acceptable during this kind of training is a matter for debate. I would only go to the point where the first tension sign is visible, then work on relaxing the horse before moving on. Using this technique, the rider will typically make progress within one session, and therefore be reassured that the training is working. To help the horse in these situations, it is very important that the rider is able to recognise the slight tension signs, know how to respond to them, and be convinced that this is worth doing! So much of my focus will be on helping the rider towards working independently on this and coming up with a solution with the rider that they are comfortable with and will actually carry out.

Avoid the unwanted behaviour

Avoiding unwanted behaviour is largely dealt with by the areas covered already; aiming to keep both horse and rider calm and comfortable, and expanding their comfort zones gradually. However, I usually discuss this idea directly with riders at some stage, as it is common for riders to have learned to do exactly the opposite; to bravely confront the issue, believing that it will not be resolved unless it is tackled head-on. Many are very relieved to avoid this! Others will struggle with the idea that progress can be made while avoiding the issue. I've generally found that the most effective way to approach this is to show the rider a way of helping the horse, then explaining why this works once they have seen and felt that it does work.

We can discuss this in as much depth as the rider wishes – the effects on overall relationship, the possible consequences of escalating aversive stimuli and punishment, the safety and welfare issues around stimulating more extreme behaviours from the horse, and so on. Usually we at least talk about the different feel when horse and rider cooperate to solve a problem, rather than the rider "making" the horse do something she is uncomfortable with.

Initially, Emma may have felt she wished to avoid Willow refusing to go forwards or running backwards; these are both certainly unwanted behaviours. We discuss this during our first session and she is relieved to hear that these behaviours can be addressed without being tackled head-on. After a couple of sessions working calmly in walk and trot, Emma's perspective changes, and she wants to avoid the little behaviours which indicate that Willow is feeling tense. She has moved from wanting to avoid behaviours that worry her to wanting to also avoid behaviours that worry Willow. We focus on encouraging more of the calm cooperation we are getting, rather than dwelling on the behaviours we don't want! When things don't go quite to plan, and we do get an unwanted response from Willow, we discuss what caused this and how we could approach things differently.

Attention

Most riders will notice a horse's attention is not on them when the horse spooks or startles, or they are unresponsive to aids, but may not notice otherwise. Clearly, when a horse's attention is elsewhere, she is likely to react to stimuli from this source, and to be less aware of her rider. Being aware of the focus of her attention gives us useful information about what she may be feeling and what she may do next. For safety and enabling quiet communication between horse and rider, we need to know how to bring her attention back to us. I would discuss the following with the rider.

- Why is it important to be aware of where your horse's attention is? Nothing works well if you don't have your horse's attention!
- How can we tell that your horse is paying attention to you? (Ears are very important here!)
- Why is the horse's attention where it is?

For example, she may be focussed on a worrying object in the arena, or another horse. If she either doesn't understand the stimuli (aids) from the rider or is uncomfortable in what is being asked of her, she may lose attention. She may be feeling physical discomfort. She may be focussed on the rider, but in a negative way, if she is worried by the rider's actions.

- How do we ask the horse for her attention? Broadly speaking, by presenting the most salient stimuli in her environment. However, if she is very distracted this may mean we would need to do something very "big" to get her attention. For example, if her saddle doesn't fit and is causing her discomfort, this is bound to be distracting and make it hard for her to calmly pay attention. Clearly, we want to resolve this issue rather than making our stimuli bigger. Discussing this with the rider, we see that starting from our safe place is the first step; where our horse is calm and comfortable. We can then talk about what kind of stimuli will get her attention. Actions which cause physical discomfort or fear will certainly grab attention, but not in a good way! Rewarding her is a good way of gaining her attention; giving a treat is one possibility, but also scratching, stroking or a calming voice – each horse is individual in this. Asking for something that she understands and is comfortable doing, particularly a gentle bend or lateral movement can be a good way of re-engaging attention. With some horses, more mental stimulation will be beneficial. So, there are no hard and fast rules here, other than that we patiently ask for the horse's attention rather than demanding it!

If you routinely ask for your horse's attention when you are riding, and the consequences of their paying attention are good, then it becomes very easy to get their attention in a calm and productive way.

During our first lesson, Emma comments that Willow has a very short attention span. Actually, Willow is very good at paying attention to some things! In this situation, Willow has a lot of attention on the "scary corner" and trying to make her pay attention to the rider's aids instead would just increase her worry. Once Willow and Emma have established a safe place to work in, Willow can relax and listen to quiet communications from Emma.

Learning theory

Nicole Graham SEBC PTC, when asked to list three key things riders should know about equine behaviour (Nicole Graham, Stirling, 2017, personal communication), replied:

1. A sound knowledge of the ABC process of learning theory.
2. Knowledge of body language and the skill to spot signs of tension. This will help them to determine what each horse finds rewarding or punishing and therefore helps them to accurately understand the consequences part of the ABC of behaviour.
3. A knowledge of how antecedents influence behaviour and the ability to learn. This helps riders to empathise with their horse and predict their behaviour which keeps them safe!!

Understanding the basics of learning theory is indeed very important for any rider. As Nicole says, learning theory on its own is not enough. We recognise that good coaching of people relies heavily on listening to the learner and adapting according to their feedback. The same applies when we train (and therefore when we ride) any horse; a knowledge of learning theory must be combined with a knowledge of body language and a willingness to listen to the horse and adapt according to their reactions. In most cases, I would keep information about learning theory fairly non-technical in a riding lesson. I would introduce it not as an abstract concept, but through the first practical example that we come across.

Giving riders the skills to avoid unwanted behaviours and therefore avoid situations where punishment or escalating negative reinforcement might be applied, is important, and will introduce some aspects of learning theory as mentioned previously. Where appropriate, I would use positive reinforcement, often in the form of scratches, sometimes a food reward, to help to shape a desired behaviour, such as standing still at the mounting block. We would usually also discuss the effect that such positive interactions have on the overall relationship between horse and rider, and on the horse's attitude towards ridden work (Sankey *et al.*, 2010).

Another area which falls under learning theory is one that good riding instructors often cover and understand well: do less. If the horse isn't responding to an aid, a common response from the rider is to do more; increase the existing aid or add another. Instead, we can pause and look at how many stimuli are already being presented to the horse and see how we can simplify things. In an ideal world, we present a single stimulus, to which the horse has learned the desired response, and the desired response is maintained by a schedule of reinforcement. In most cases, as shown in the example below, doing less so that the stimulus (aid) can be quiet and still noticeable to the horse is a much better solution. Once the rider has applied learning theory in a couple of situations, we can help them to generalise the concepts so that they can problem solve for themselves more effectively.

I remind the rider as necessary to keep listening to the horse as they develop their skills. Without this monitoring, it is easy to be guided by our beliefs as to what is aversive or appetitive to the horse in each moment when it is the *horse's* perception that matters. The *correct* application of learning theory depends on an accurate reading of the horse's emotional response to stimuli; for example, if we believe a hearty slap on the neck is a reward and the horse will learn to repeat behaviours which are followed by this slap, we may not get very far! The *ethical* application of learning theory also depends on this awareness of the horse's emotional response. It is possible to apply learning theory correctly, in that the desired behaviour is reinforced, but in a way that compromises the horse's welfare. This will be illustrated in the example below.

I often refer interested riders to Frans de Waal's TED talk on moral behaviour in animals (de Waal, 2011), which discusses interesting research that might help us to think beyond immediate consequences to the more far-reaching effects of our interactions with horses.

The basics of learning theory are important but can be a rather blunt instrument! I personally feel we are aiming to ride in such a way that the horse enjoys being ridden, as opposed to the horse tolerating being ridden in return for subsequent rewards. The horse should at least find ridden work engaging enough that big external motivators, be they positive or negative, are not required. Some will feel this is an unrealistic goal, and for many horses it will be; if they have physical issues that make any riding uncomfortable, for example.

For a horse in good physical condition, there are many aspects of riding that they might find enjoyable. For example, coordinating movement with a person they feel comfortable with; the mental stimulation of problem-solving; as well as some of the movements themselves. So, along with the basics of learning theory, I'd often discuss motivation with riders, distinguishing between:

- external motivation where the horse learns that life is better if they respond to a given aid in a certain way; the uncomfortable pressure on their side stops or they receive a reward for their response;

and

- internal motivation, where they are engaged by what they are doing, rather than focusing on the consequences of performing a given action.

We are looking for the horse to enjoy their work, rather than needing to be coerced into doing it by being punished or paid. We would like the horse to work *with* us rather than *for* us. In many cases this might lead the rider to reconsider the kind of ridden work they do with their horse; do they enjoy flatwork or jumping, hacking out alone or in company, a wide variety of work and new challenges? This is by no means a new idea – trainers for centuries have considered what the horse shows enthusiasm for when planning their training.

> Returning to Emma and Willow, initially Willow was faced with a number of aversive stimuli when she entered the riding arena. Misunderstandings arose between Willow and Emma. From Emma's point of view, Willow was "getting the better of her". From Willow's point of view, events like being kicked, smacked with a schooling whip, a strong and rather tense contact on the reins and approaching the "scary corner" probably all registered as aversive. She was reluctant to move forwards at all.
>
> Let's say we would like Willow to move forwards in response to a very light leg aid, just a brush of a leg against her coat. We could crudely apply learning theory to this issue in several ways. We could use negative reinforcement, applying an aversive stimulus until she moves. We may achieve the desired behaviour in this manner, but at the cost of Willow's emotional well-being, as she is now having to cope with more aversive stimuli. Or we could use positive reinforcement, for example giving her a scratch or a treat when she moves. If we gave a big enough treat for moving, and shaped her behaviour skilfully, we could get her moving forwards without changing Emma's riding at all. However, if moving forwards still feels somewhat uncomfortable, then she will only continue to do so while she is getting "paid" for the movement, and we are combining aversive stimuli with rewards in a way that may well lead to conflicting motivations for Willow, thereby causing tension and unpredictable behaviour.
>
> It is far preferable to address this issue as we did, by changing Emma's position and movement so that moving forwards is comfortable for Willow, and even enjoyable. We remove as many aversive stimuli as we can, and then evaluate Willow's attitude to her work, keeping in mind that she may still simply find it disagreeable for reasons we can't address!

So, whilst we use learning theory to shape the horse's behaviour, we keep within certain boundaries depending on the horse's response. It is important that the horse is not worried enough to be considering avoidance, and that minimal coercion is used, be it positive or negative. In short, we are applying the LIMA principle (introduced in Chapter 1) to our riding.

Control

The conventional interpretation of control within the context of a riding lesson would be the rider having full control over the horse, giving aids that the horse understands and responds to as the rider wishes. In a riding lesson delivered with behaviour in mind, what might we change about this and why?

For both the safety of the rider and the welfare of the horse, ensuring the horse has control over aversive events is important. There is a substantial body of research showing that control has a significant influence on both psychological and physical well-being. In particular, control over aversive events has been shown to render them far less stressful than aversive events over which the individual has no control (McMillan, 2005). Importantly, Robert Sapolsky (Sapolsky, 2004) notes that the exercise of control is not critical; it is the belief that you have control which reduces stress and associated damage. To take an example we can all relate to, consider a trip to the dentist. If the dentist tells you that she'll stop if you raise your hand, most people will believe they have control, and will find the situation less worrying than if the dentist were to secure you to the chair and make it clear they would carry on regardless!

So, we aim to minimise aversive stimuli, and ensure that the horse has some control over them when they do occur. We shape the rider's behaviour and develop her skills so that her aids are not aversive, but part of a calm two-way discussion between horse and rider. Along the way, we ensure the horse gains more control over aversive events.

After several sessions with Willow and Emma, Willow has much more control over aversive events. Emma still reverts to too strong a leg or rein aid on occasion, but she does immediately remove the pressure when Willow responds as Emma wishes – so Willow knows what to do to remove the aversive stimulus. As Emma's skills improve, she will present fewer and fewer aversive stimuli to Willow.

Looking at control in a slightly different way, if we aim for submission in the horse, this may imply that the horse should behave as the rider desires regardless of how she feels about it. If, however, we are aiming for cooperation, then both parties must have some degree of control. So, something I might discuss with the rider is how she feels about letting the horse say no. When the rider gives an aid, if the horse doesn't give the desired response, there are options rather than simply repeating the aid or making it stronger, which both lead back to an adversarial form of riding. We could instead think about how we asked the horse, our timing, whether the horse understands the aid, whether we are being clear enough, whether the desired response is physically possible at that moment in time, and so on. In broader terms, we might be willing to accept that there are just some activities this horse finds unpleasant, however careful we are with our training, such as hacking out alone. Some horses seem truly comfortable going out with just their rider for company; others may never relax and be comfortable with this. Interestingly, my experience has been that if you take this approach, the horse will become more cooperative and responsive, and on the rare occasions where you do require a certain behaviour in a timely fashion (for example, if the horse must move in a certain way to avoid traffic), they are much more likely to respond well than the horse who is forced. This approach leads to big improvements in the rider's skills. Rather than simply learning to "keep kicking" or "kick harder", the rider can really take responsibility for working out whether what she is asking is reasonable from the horse's point of view, and how to ask in a way that will get a calm, cooperative response.

How do you teach riding with behaviour in mind?

So, we have some idea of what we'd like to teach about behaviour in a riding lesson. And we can certainly give lots of reasons supporting this approach, as covered throughout this book. However, what we teach might be quite different from what the rider is used to. As well as developing physical skills, we may well be trying to help them develop and change attitudes, beliefs, emotional responses, to develop more self-awareness and awareness of the horse as an emotional being — and do a lot more thinking! This can be fascinating, or it can seem like a lot of hard work. If we, in our great enthusiasm and desire to "convert" riders, flood them with information about this approach, we run the risk of it all seeming like a lot of hard work, or conflicting so much with the rider's current beliefs and habits that they cannot accept these new ideas.

This isn't the rider's fault, it is perfectly normal human behaviour. To give a common example, have you or someone close to you ever tried to lose weight? You can have lots of motivation; to look better, feel better, possibly live longer, fit into clothes you can no longer wear, and so on. And you can have lots of information both about the benefits and how to go about it. Does this make it easy? Not at all. And if the process feels too hard, your brain can be amazingly creative in coming up with reasons not to act. So, our job is to keep at the fascinating end of the scale as much as possible and introduce changes in a way that doesn't seem overwhelming or like too much hard work. How we work with the rider is therefore extremely important; it is a huge topic, and very individual to each combination of rider, horse and instructor. Maybe the simplest way to start riding instruction with behaviour in mind is by integrating behavioural goals into our coaching plans.

Going back to Willow as an example, we could express a goal to change Willow's behaviour of avoiding one corner of the arena in several ways. At one extreme, we could say our goal is to "Make Willow go into the top left corner of the arena". This statement is confrontational and may well imply the use of force if necessary. All we are focussed on is Willow moving to that area of the school. The rider may well feel they have failed if they can't make this happen, and we're not considering Willow's welfare at all in this goal.

At the other extreme, we could say that our goal was to "Help Willow overcome her fear of approaching the top left corner of the school." This creates a totally different way of thinking about the issue. Helping Willow implies cooperation rather than confrontation, and our focus is on reducing her fear. The training plan that will be developed based on this goal will necessarily include the elements we have discussed in the previous section; avoiding the unwanted behaviour, awareness of body language, changing attitudes, control, and so on.

Another interesting question is when and how to set coaching goals. Drifting on without any structure won't work for too long, but neither, generally, will setting goals very early on. If the rider is, for example, still very focussed on "making" the horse achieve something, and doesn't have knowledge about the areas we have discussed above, then it is unlikely that a plan that both instructor and rider can feel enthused about will be devised. So, giving the rider some new information to base her decisions on is important.

The method I most often employ initially, if a rider is new to this kind of lesson, would be to assess horse and rider (only up to a point where I think safety and welfare are not compromised), asking the rider questions about herself and her horse. I would then suggest that we experiment with some changes and ask the rider how they feel and how they think the horse feels. As with any other coaching, I would start with clear ideas of where my boundaries are, in terms of what I feel is OK for horse and rider and explain these bit by bit to the rider, gradually handing over responsibility for assessment and problem-solving to the rider. This method was followed in the examples of Emma and Willow, and the examples given below. This is again not so different from regular good coaching practice. If your client comes in wanting to jump a course of jumps, but is very unbalanced in trot, you will use a similar approach to try and adjust the client's expectations whilst keeping her involved and motivated.

In addition to careful definition of goals, I find thinking of training the rider in much the same as you would train the horse very helpful; avoiding the unwanted behaviour in the rider, awareness of body language, using slow shaping, using LIMA, and so on (see chapter 3). This process is illustrated by the example of Emma and Willow running throughout this chapter; a couple of extra examples may also be helpful:

Example 1

Linda has called me because she is having problems with Blue, a 9-year-old thoroughbred. He is rearing in the showjumping warm-up arena; he is also rearing and spinning if she tries to ride him out from her home. Linda is passionate about showjumping, and early on in our discussion she says she must be able to compete with Blue – immediately, and at least once a week – so it is very important that this problem is resolved. She says she knows she needs to make him stop being so naughty, but her riding instructor has recommended unbalancing him when he rears to put him off doing it, and Linda is worried about her safety if she does this. I simply say that I understand what she wants, and we'll look at her and Blue and see what is going on. As we progress with our first ridden session, Linda feels safe with Blue for the first time in months. Initially, I keep a fairly strict control of what is happening; explaining what I am seeing and helping Linda and Blue to both relax and helping Linda to recognise tension in Blue and help him. After half an hour of work in an arena, they have made great progress, and we go out onto the track where Blue usually rears and spins, being very observant of tension signs in either Linda or Blue. We go much further down the track than I expected; as we approach the place where Blue usually stops, he still looks relaxed. We stop here briefly to talk, and unasked, Blue takes a step beyond where he would usually refuse to go forwards. The step itself looks tentative, as if he expects the ground to give way. But then his confidence seems to return. Both Linda and I are genuinely surprised and delighted with him. Linda can feel that this was a big change. On the way back up the track, she says "that's what I want most; that feeling of Blue and I working together, it was amazing". At this point, we start to discuss goals and a training plan. Linda is a very competent and experienced rider, and only a few lessons are needed to help her and Blue through this issue.

Key points
- *Setting goals*: Initially, Linda expressed a strong desire to get Blue out competing

regularly as soon as possible – without the rearing. By delaying discussion of this until Linda had some new information to work with, we could then set a goal that includes conventional activities such as competing in showjumping, but prioritises calmness and cooperation between horse and rider throughout all stages of training.

- *Changing attitudes*: I haven't tried to make Linda change her attitude towards Blue, but as her understanding of him develops, her way of thinking about him changes.
- *Avoid conflict between rider and instructor.* If I had started with any implication that making Blue stop his naughty behaviour was misguided, or unfair to Blue, either Linda would have believed me and felt awful, or she would have rejected my ideas and been reluctant to work with me. Neither would be productive. We don't want to confront our horses and provoke negative emotions; we need to also apply this to the riders. Instead, we introduce new ideas about focusing on how Blue is feeling and experimenting with how to help him relax.
- *Use slow shaping.* For example, lots of repetition is needed to remind Linda to check on Blue's emotional state. When the environment changes, we need more repetition. For example, Linda may pick up Blue's anxiety at the mounting block but forget about it when asking for a transition from walk to halt. She quickly understands that big leg aids are not helping, but still needs reminding before giving an aid.
- *Avoid the unwanted behaviour.* Initially, I always remind Linda to give a quiet leg aid before she applies it, rather than correcting her afterwards.
- *Reward the smallest of improvements.* I observe Linda carefully, and am quick to pick up on and reward small changes, such as more empathetic language when discussing Blue, or noticing mild tension.
- *Be genuinely enthusiastic about their progress.* This applies throughout the lesson; particularly at the end – Linda truly helped Blue to relax and think through taking that extra step forwards, and I was genuinely touched and impressed with both of them. This requires a level of awareness of your own feelings and learning how to change things if you begin to feel, for example, frustrated or impatient with the rider.
- *Notice what motivates them and helps them to change.* At the beginning of the lesson, Linda said she needed to be out competing regularly, and fixing this problem quickly was important. However, even as we took our time over Linda mounting Blue, I could see that his relaxation was helping her, and the biggest reward for her immediately was to not feel scared of him. Often, we motivate change initially through the rider feeling safer; when the rider is scared, it is hard for her to consider the horse's feelings; once she feels safe, her observation of and concern for the horse's emotional state can develop.
- *Look for causes of unwanted behaviour.* Linda's behaviour towards Blue was caused by a combination of fear, lack of understanding and learned behaviour, so we worked on addressing these issues rather than punishing her for her actions.
- *Set a good example!* My feelings towards both Linda and Blue will have been obvious through my body language and actions, and this will often influence the rider more strongly than words alone!

Example 2

Sarah has come to me for a riding lesson; she doesn't mention any behaviour problems.

However, as we chat beforehand, she describes her fell pony Thalassa as "headstrong, always rushing and fighting with me" and is particularly frustrated because she "won't go in an outline". Thalassa is Sarah's first horse, and she wants to compete her in dressage. Sarah reports that when she really tries hard to slow Thalassa down, Thalassa often responds by bucking, and Sarah has fallen off a few times. After ensuring that Thalassa has had her tack checked by an appropriate professional, and full physical checks, we have our first lesson. Here, our approach is very similar to that described with Emma and Willow; both horse and rider are tense and fearful of each other, so we find a safe place to begin work. We create an environment where Sarah can comfortably release her grip on the reins and see that Thalassa is much more comfortable with this. We then work on Sarah's riding, and discuss Thalassa's responses to the changes Sarah makes. We discuss the triggers for Thalassa rushing and bucking. The primary cause of this is Sarah's anxiety, which makes her tighten her whole body and pull hard on the reins. Sarah continues to have regular lessons, and they steadily expand their range of safe places.

Key points

Many are the same as those discussed above and will not be repeated here.

- *Notice what motivates them and helps them to change*: As with Linda, feeling safe is a strong motivator for Sarah. However, we are going to move slowly, and the temptation to listen to conflicting advice: that she should "be brave" and "sort it out" may be strong. Peer pressure may also come into play, as she is not cantering around the arena and will be told she is being too soft with her horse and will "ruin it"! Before we started riding, Sarah showed me some videos of ridden work. During our first session, I take some video; only in walk, but both horse and rider are moving much better, and in a "better outline" than previously. I edit a before and after video and email some comments from well-known dressage riders on the importance of self-carriage, establishing a good walk, and so on, and give advice on how to handle critics.
- *Setting goals*: We don't discuss goals until the beginning of our second lesson, giving Sarah time to reflect on the changes she felt and the videos and extra information I sent. By the start of the second lesson, Sarah understands that relaxation in both horse and rider is important both for safety and welfare, and for her desire to compete in dressage.
- *Changing attitudes*: Sarah quickly realises that Thalassa's behaviour is not in fact due to her being of bad character but triggered by specific events. She is affectionate and considerate towards Thalassa on the ground and applying this when they are riding too makes absolute sense to her. She has a good level of self-awareness, and realises that when she is feeling worried, her attitude to Thalassa changes, and she can address this by working on slowly expanding their comfort zones.
- *Avoid the unwanted behavior*: Sarah is very likely to move too quickly. Once she feels the change in Thalassa in walk, she wants to apply it to trot and canter work immediately! I need to be very clear about how much repetition is required, and how gradually their work should be expanded.
- *Reward the smallest of improvements*: Sarah needs more attention in this area than Linda. Linda can feel the changes and is very sure of their value, so confirmation from me is helpful but not essential. Sarah

is much less certain and needs more guidance, encouragement and explanation.

Final thoughts

Throughout this chapter, the examples have been of riders with their own horses. What about teaching clients at riding schools? Whether in a riding school or with a private client, I would initially want lessons to start at first contact with the horse; catching, leading, grooming, tacking up. All the points throughout this chapter apply to the whole process; the rider is always encouraged to think about the horse's point of view and be considerate and observant. Working with the horse on the ground is a good opportunity to help the rider understand her better and feel more empathy for her, as well as improving problem-solving skills. The lesson ends when the horse is untacked, groomed as required and returned to her living area. This is common practice in good traditional riding schools.

So, a lesson with a single rider in a riding school will be much the same as that with a private client. In a group lesson, things may be a little different. Riding instruction with behaviour in mind is very learner-centred, with the focus on two learners: the horse and the rider. So, explanations, exercises and overall goals are very much adapted to suit the individuals in that moment. We can still set general goals for a lesson but need to be aware that every pair will be starting from a different point and have different needs, so we won't ask for the same things.

For example, let's say we are working on transitions. We might set an exercise, but rather than saying that each rider should transition between walk and trot at each letter of the school, we would make calm transitions with minimal aids the priority. We would then help each rider to assess a reasonable starting point. So, some combinations might be able to work on trot/walk transitions at each letter, others might work on walk/halt transitions down one side of the school. We improve riders' understanding of what is needed for a calm and cooperative transition; did the rider have the horse's attention, was the horse balanced enough to manage the transition smoothly, did the horse understand the aid, did the rider "try too hard" and tense up a bit, or give some unintentional aid? Can the rider feel when the transition happens smoothly and calmly? Do they notice the horse's head popping up when they give slightly too big a leg aid? And so on. We work towards smooth, calm and timely transitions, with the focus on doing this well rather than just making it happen. This can be made more interesting by challenging riders to see how little they can do to get the desired response, for example, and getting observers to notice what aids were given, the goal being to keep aids invisible!

In a jumping lesson, a variety of jumps could be offered, and the emphasis would not be on getting over the jump, but on choosing the appropriate jump and not taking it if horse and rider aren't able to do so in a calm and balanced fashion.

Finally, it is worth considering the responsibilities of a riding instructor. Behaviourists will view themselves as part of a team who look after the horse's well-being and training. This team will include vets, farriers, saddle fitters, qualified physical therapists, riding instructors, and so on. It is essential to stay within your area of expertise for the welfare of the horse. If, as a behaviourist, you suspect there is a physical problem, you refer to a vet rather than giving an opinion on what is wrong. If you are a behaviourist who is not a riding instructor, you refer relevant aspects of ridden problems to a riding instructor and work with them. Likewise, if you are a riding instructor with an interest in behaviour, but it is not your area of expertise, it is important to refer to an accredited equine behaviourist rather than experimenting with solutions to see if they work. As with physical problems, behavioural problems require diagnosis and treatment based on this diagnosis and working with a behaviourist can be very beneficial for all concerned. Riding instructors are often expected to be the source of all knowledge about every aspect of the horse, but recognising your area of expertise, sticking to it and referring to other professionals as appropriate keeps everyone safer, happier, and is a great way of learning more about these fascinating beings – none of us know it all!

References

Fédération Equestre Internationale (2017) *FEI Dressage Rules*. 25th edition. Fédération Equestre Internationale, Lausanne, Switzerland.

McMillan, F.D. (2005) Do Animals Experience True Happiness? In McMillan, F.D. (Ed.), *Mental Health and Well-Being in Animals*. Blackwell Publishing, Oxford, UK. pp. 221–233.

Oliveira, N. (1988) *Reflections on Equestrian Art*. Translated by Phyllis Field. J. A. Allen & Co. Ltd, London, UK.

Overall, K.L. (2005) Mental Illness in Animals – The Need for Precision in Terminology and Diagnostic Criteria. In: McMillan, F.D. (Ed.), *Mental Health and Well-Being in Animals*. Blackwell Publishing, Oxford, Great Britain. pp. 127–143.

Sankey, C., Richard-Yris, M., Henry, S., Fureix, C., Nassur, F. and Hausberger, M. (2010) Reinforcement as a mediator of the perception of humans by horses (Equus caballus). *Animal Cognition,* 13: 753–764.

Sapolsky, R.M. (2004) *Why Zebras Don't Get Ulcers: The Acclaimed Guide to Stress, Stress-Related Diseases, and Coping*. 3rd edn. Henry Holt & Co., New York City, New York.

Swift, S. (1985) *Centred Riding*. 1st Edn. William Heinemann, London, UK.

de Waal, F. (2011) Moral Behaviour in Animals. Available at: https://www.ted.com/talks/frans_de_waal_do_animals_have_morals (accessed 26th February 2018).

Wood, P. (2012) *Dressage the Light Way*. 1st edn. J. A. Allen & Co. Ltd, London, UK.

Xenophon (1999) *The Art of Horsemanship*. Translated by Morris H. Morgan. J. A. Allen & Co. Ltd, London, UK.

Chapter 7

Rehabilitation and rescue

Ben Hart

Whether undertaken privately on a small scale or professionally on a larger scale, a behaviour-minded approach to rehabilitation and rescue would be pertinent. For example, rescue centres often have to deal with regular introductions of horses to each other, which can cause injuries through kicks and bites, or social deprivation if not attempted at all. A behaviourally-minded approach to introducing horses will be discussed in this chapter along with how to "solve" some problems required for rehabilitation of rescued animals.

While it is easy to think of rescue and rehoming as something only undertaken by organisations whose specific remit is to provide care for the benefit of equines, it is important also to acknowledge that rescue is often undertaken by individuals or small groups of well-meaning equine enthusiasts. Therefore, in this chapter I link together these different groups of people and also keep in mind that when individuals take on loan or rehome a horse from an equine charity, they are continuing the rescue process. The key concepts of a behaviourally-minded approach can be applied to the rescue and rehabilitation of equines in relation to a large organisation, the individual taking on a rescued equine and situations in-between. For ease of communication throughout this chapter we will use the term "rescue centre" and intend that this term is all-encompassing from large organisations to individuals.

Park the physical

The first thing to say is let's "park" the physical. As is abundantly clear from other chapters of this book, physical pain and veterinary conditions have a major influence on the behaviour of equines. While writing about rescue and rehabilitation, it is impossible to include every scenario of physical discomfort or pain due to a medical condition that might influence the behaviour of the animals involved. It goes without saying that every effort should be made to remove or reduce physical discomfort and any conditions that might influence the behaviour of the animal. That said, it is almost impossible to separate out behaviour and physical conditions when an animal is first entering the rescue situation, therefore it is my intention to discuss how our behaviour towards the animal during these initial contacts can be hugely influential in the animal's acceptance of future handling. However, for the sake of clarity and focus, let's assume the organisations involved in rescue have removed physical pain and discomfort before further embarking on the animal's rehabilitation and training.

A pony called Pocket

Pocket was a dappled grey pony who had found herself in a rescue centre after bucking off her young owner one too many times. She had also become "aggressive" and sadly the owners had not sought professional help but instead neglected the pony. For the year before she was eventually seized by a rescue organisation, she had lived her days in a dirty stable on a ramshackle livery yard – the sign above her stable read "Bites, do not approach". Her feet were overgrown, her back pain (the cause of the bucking, still untreated and her outlook bleak. Upon rescue, handled by experienced individuals with behaviour in mind, she entered her new life. At the small and well-run rescue centre she was turned out in a herd, after an appropriate introduction process, and allowed to be a horse again. She was then gradually handled to accept routine procedures and her back issue was addressed by a course of veterinary and physiotherapy treatment. She was brought back into work by the grooms who introduced her to ridden work under the guidance of a behaviourally-minded riding coach and in time she was considered ready for homing. She had some challenges to overcome with respect to living with other horses – we will come back to her story later.

The first few days

Most animals that enter the rescue system do so either because of physical conditions, which have often led to behavioural issues, or because the animal has behavioural issues. Animals that are relinquished to rescue centres and do not have any significant veterinary conditions, are easy to handle and have no behaviour issues, are unlikely to stay long in the rescue centre as they are easy to rehome. Their fast exit from the centres allows valuable resources to be used for other horses in need and those that require more support. For this reason, rescue is inevitably focussed on dealing with behaviour: issues must be solved primarily to facilitate optimum care of the animal but also to ensure the safety of the staff and to maximise the opportunities for rehoming.

One of the most common challenges facing any rescue situation is the initial handling and treatment of equines that are coming into their care who have had minimal or no handling; how this can be done safely for the animal and for the humans involved. To put it another way, what do we need to do when we just have to get it done?

This initial contact with the animal is the most dangerous both for the animal and for the handlers, both in terms of their physical safety and in the sense of the animal's long-term view of their new environment and their interaction with humans. The appropriate use of physical restraint and veterinary intervention, and the use of licensed drugs for sedation, is the key for safety and ongoing improvement in the animal's behaviour. The ability of those handling the animal to understand behaviour and use appropriate techniques to minimise the fear and negative experiences the animal is likely to experience in these situations, is crucial.

During the animal's journey through the rescue and rehabilitation process, it becomes obvious that it is not what is done to the animal, but how it is done that is the determining factor of the animal's future rehabilitation. This is solely down to the knowledge and practical application of the science of behaviour not only by those people involved in handling the individual animals but also those planning and creating the environment and protocols for the rescue centre.

Giving the good experience

Let me give you an example of both the complexities of working with behaviour in the unhandled equine and the importance of understanding how the animal's experiences can influence their later behaviour. In 2015, while attending the Donkey Welfare Symposium for The Donkey Sanctuary at the University of California, Davis, California, USA I had the pleasure of working with 20 Bureau of Land Management Burros (donkeys). These animals had once been free roaming on areas of Californian rangelands and had been removed as part of the management of these areas. Upon removal from the range, the donkeys are kept in large government-owned corrals. At that point, they have had minimal handling, not been inoculated and their only close experience with humans has been when they were branded for identification purposes and when they were herded into squeeze chutes for their feet to be trimmed.

On arrival at the holding yards for the symposium, closer inspection revealed that one of the jennies (a jenny is a female donkey) had a wound in the heel of an off hind leg, deep enough to be clearly seen and painful enough to cause her to limp. So the scenario that presented itself was an unhandled donkey with a general negative association with people and a wound that needed veterinary intervention. In addition, the environment was not set up for handling injured equines, just a standard stable and holding yard for each group of two or three donkeys. The only thing in our favour

Figure 7.1: A donkey from the Bureau of Land Management being treated on day one of Donkey Welfare Symposium. Photo: Ben Hart.

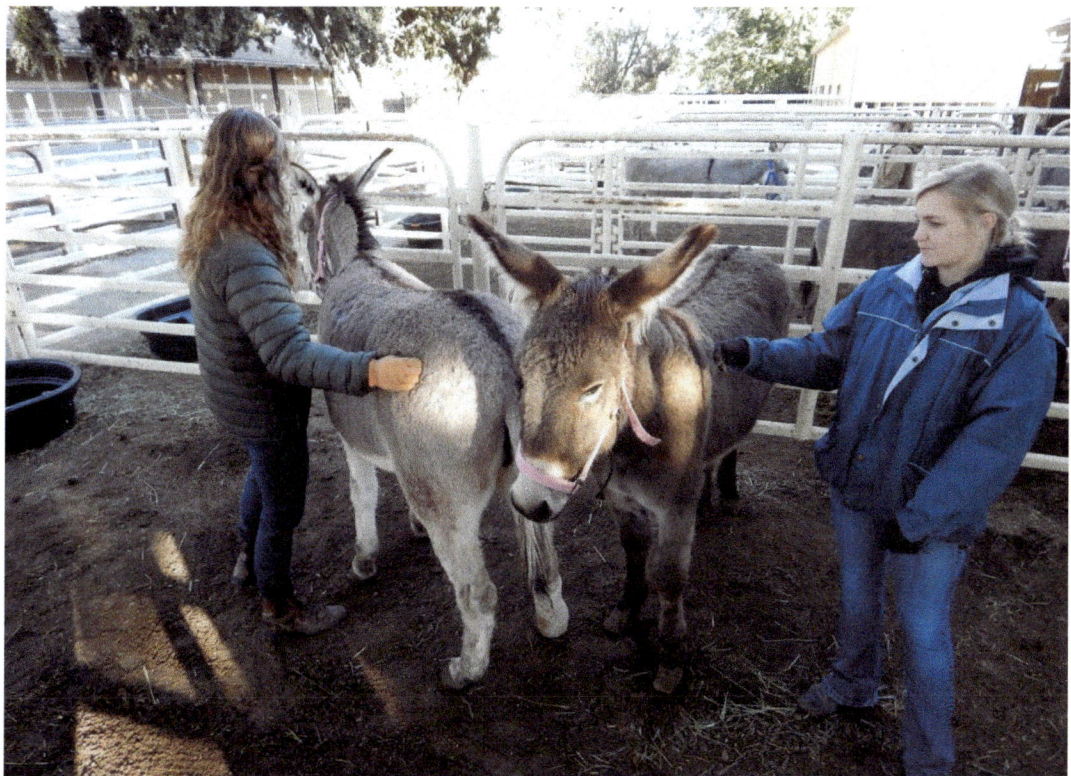

Figure 7.2: The same donkey on day 3. Photo: Ben Hart.

was that the Bureau of Land Management had run the donkeys through their handling system before loading and transportation and placed a head collar on each animal.

The starting point for restraint in this situation is not the physical conditions of being able to safely immobilise the animal, it is a team that starts with a desire to "give the good experience"; if all members of the team come to the process of having to "get the job done" with the desire to give the animal the best possible experience, then that informs the way that the team will interact with the animal and each other. The mistake rescue centres often make in these situations is that the later rehabilitation of the animal is seen in isolation from these initial interactions with the animal and their handlers. Often, this is because the early focus for the centre is on treatment and physical conditions rather than on behaviour, which is understandable but nevertheless likely to be detrimental to the animal. Every rescue centre should be mindful that the rehabilitation process starts from day one and inappropriate or stressful handling that occurs at this early stage will be detrimental to the animal's progression through the rest of the rehabilitation process.

Using negative reinforcement in the form of mild pressure and release to approach this jenny, I was able, after 15 minutes, to shape her behaviour for catching, without the typical "lunge and grab". Then I prepared her, again through shaping, for the process of restraint. Restraint of an equine could form an entire chapter on its own and each rescue centre should consider its restraint policy very carefully before implementation. Primarily, it's important to ensure that the handlers are physically capable of safely restraining the animal and that their knowledge of behaviour allows correct interaction based on scientific principles. The key factor at this stage for this jenny was that the process was being controlled by a behaviourist rather than any other member of the team and this allowed for time to be removed as an influencing issue in the way that the mare was handled; we'll discuss this later in the chapter.

Understanding the equine's likely reaction to being restrained allows the animal to be calmly immobilised in preparation for intravenous sedation. During the restraint process, it is normal for equines to struggle as they establish whether they are truly restrained or whether there are escape possibilities. The insistence that handlers restraining any point must be physically capable of doing so or that the system of handling must ensure that the animal is completely restrained, ensures that the animal will very quickly give up their attempts to escape and will stand calmly.

Through experience of handling and administering drugs to unhandled equines and through the use of systematic desensitisation, the vet prepared the injection site and was able to calmly administer sedatives to a level that produced recumbency, total anaesthesia that allowed the heel to be examined, cleaned and treated, antibiotics administered and the wound site dressed.

Over the next four days, this donkey required two further restraint sessions and the use of sedation to allow continued treatment of the wound. However, what is particularly interesting is that this jenny, despite this unpleasant intervention, progressed the furthest with her handling of all twenty donkeys. By the end of four days' training, she actually sought out human interaction and showed a clear desire for physical contact – scratches on both the withers and rump became reinforcing for her.

I hope that this story highlights the fact that having to get something done in terms of treatment does not mean that the animal's behaviour has to deteriorate as a result of such intervention. Provided that the team involved in treatment has the desire to give a good experience, equines are incredibly forgiving and it must be remembered that they are learning in all situations not just during deliberate training sessions. I consider there to be three elements that come together in any rescue centre to give the good experience; the environment, the humans involved and the training methods the humans use.

The environmental element

As covered in other chapters, meeting the needs of the equine is crucially important to their welfare. The better the environment in terms of allowing the animal the opportunity to be calm, rested and have a fulfilling experience, the less frustration, anxiety and stress they bring to the human and training elements.

Often, the environment is one of the most challenging aspects for a rescue centre to get right. Most centres require a period of isolation in a new arrivals unit to prevent possible spread of disease alongside veterinary interventions to establish any threats to both the individual animal's health and that of the herd they may be joining. This is potentially the most stressful period of the animal's time in the rescue centre. It must be acknowledged that it might not be possible to meet all the needs of equines under the restrictions of typical isolation procedures (e.g. direct social contact with other animals and large acreages of good grazing). However, it is possible for rescue centres to create behaviour-minded procedures for new arrivals that meet their needs as much as possible within the necessary limitations.

Diet

Perhaps the most basic of environmental needs is a diet of high-fibre forage available on an *ad lib* basis without conflict with other companions. Several water sources should also be provided.

Movement

Ensuring as much opportunity to move as is practically possible is crucial to the well-being of an animal in this early stage of their rehabilitation. Along with movement comes space and that will include sufficient space to roll, to feed, to lie down and to rest comfortably without social interference from other animals.

Social

The companions, including humans, which the equines find themselves with will influence their behaviour during your arrivals procedures. Ensuring that individual animals are kept in their bonded pairs is especially important in donkeys who can become stressed when separated from their companions, which in turn has a possibility of leading to potentially life-threatening conditions such as hyperlipaemia. Lone animals should also be able to see other equines, but during the new arrivals process it should be ensured that new animals are not intimidated by being in close proximity to unfamiliar animals. They must also be able to relax, lie down and rest uninhibited or intimidated by others.

Simulation

Providing an enriched environment in which the animal has the opportunity to explore and perform natural behaviours such as chewing or playing with objects, will help fill the animal's day and prevent boredom and stress.

If you can get the environment right, you will have a calmer and more relaxed animal than you would otherwise. However, by the very nature of the work involved in rescue and rehabilitation, centres are increasingly under pressure to take in as many animals as possible. Consequently, many rescue centres are extremely short of resources to provide ideal environments for all their residents. Overcrowding, lack of turnout during winter months, competition for food, space, limited mental stimulation and no environment enrichment, all contribute to the challenges that animals face in their new situation.

After medical conditions, environmental stress is a huge influence on the behaviour of equines. If we play with this concept for a moment, it's plain to see that when presented with an animal who has behavioural difficulties, one would be unlikely to recommend an environment that contained several unfamiliar animals that also were likely to have behavioural issues, combined with limited opportunities to perform natural behaviours, alongside limited staffing and pressure on staff to rehabilitate them as quickly as possible. However, it is a situation most rescue centres find themselves in, and although it is unrealistic to expect every centre to be able to provide for the animals' needs as comprehensively as we might do in an individual home, anything and everything that can be done to increase the way in which the animals' environment meets their natural needs will contribute to the animals' rehabilitation.

One of the biggest considerations for the management of a rescue centre might be to reduce the number of spaces available to new cases: reducing the stocking density and providing a more suitable environment for those animals that are in the centre's care would likely improve the experiences of the animals that are taken in. This is an ethical dilemma because we must question whether rescue and rehabilitation is about sheer numbers of animals going through the system or the quality of the experience that those animals have. Potentially, by having an environment that meets their needs more fully, their rehabilitation might be quicker and therefore there might be a greater throughput of animals, which in some way will compensate for holding fewer animals at any one time.

The human element

Most humans have two goals when they work with rescued equines: to keep themselves and their animals safe and for the animals to be "happy". Therefore, it is important to have these goals in mind when working within the rescue environment. Safety is paramount for both the humans and the animals. If a member of staff becomes injured, there is an impact on the rescue operation to provide care for the existing animals. If an animal causes an injury to staff, they are liable to be labelled as difficult, dangerous or beyond help, which has a serious impact on their future.

One factor that consistently undermines safety is the issue of time. It has been my experience all over the world in every possible type of rescue centre that one of the most common statements I hear from staff is "We don't have time". Time is a crucial element in maintaining safety and creating a relaxed animal and every organisation that works with equine behaviour has to find a way of "creating time" – the perception of not having a deadline on behaviour is a key principle for good care and rehabilitation.

Most accidents happen when equines are fearful, rushed, under pressure, or not given sufficient time to calm down – all a result of staff taking shortcuts in handling and training.

Whether in a commercial operation, or a charity or not for profit organisation, it is important that staff work efficiently to get the best use of the resources available. However, embedding the value that people must take time wherever and whenever needed to ensure the safety of both themselves and the animal and to provide the optimal good experience is crucially important to the success of the whole operation.

Often, staff in rescue centres feel pressured to process animals through the organisation and into rehabilitation and onward to adoption or loan homes in order to create spaces for more animals to enter the system. It is easy to see where this pressure has some legitimacy, as an organisation having space to offer to the neediest of new cases is socially important and provides great fundraising opportunities. (And before we condemn the fundraisers' motives, it is important to remember that without fundraising there is a very limited supply of financial support to provide for any of the animals within an organisation's care.) However, perceived pressure to get animals out more rapidly than the staff might think is in the animal's best interests can impact on staff morale and safety. If animals are pushed through the rehabilitation process too quickly in order to be able to be rehomed, the vital question is whether they will stay in that home. If their training and rehabilitation has been rushed, they might show behaviour problems in the new home, requiring either more support from the organisation or the animal returning to the centre – the associated costs with failed rehoming are significant both in terms of financial cost to the organisation and stress to the animal.

A scientific approach to training

One of the biggest challenges facing any rescue centre in terms of the human element is ensuring that all staff approach the rehabilitation of equines in the same manner. With so many different training methods now available and advocated by so many equine professionals, often members of a team have completely different ideas of how an equine should be trained and how unwanted behaviours should be tackled. This leads to problems with continuity and consistency, not to mention some conflict within the team itself.

If a member of staff is working in a particular way with a rescue animal, it can be demoralising for that member of staff and confusing for the equine if somebody else within the team handles the animal in a different way. Obviously, equines can learn to associate different people with different handling methods and techniques, and to some degree it could be argued to be necessary in order to prepare the animal more fully for rehabilitation and rehoming. However, if introduced too early in the training, inconsistencies and conflicts that are set up between different training methods can be detrimental to the animal's progress. One of the challenges that is faced by any rescue centre is to ensure that all the staff are working to the same guidelines and with the same processes and principles that will ensure consistency and continuity throughout the rehabilitation process.

The challenge of getting all the team working in the same way is easier in small centres or when an individual is involved in rescue or rehabilitation; the larger the organisation and the number of team members that may be involved in the rehabilitation of individual equines, the harder it is to provide consistency of approach.

As a general approach, I would always recommend that each animal is initially only trained by one person; this of course might mean a particular member of staff may have several animals to train. Consider one person as the lead trainer for a particular animal, with other members of the team following up with the generalisation work and ensuring that the animals are used to being handled safely by different people, which is needed for animals to be considered ready for rehoming. Later, we'll look at the use of shaping plans to assist with this process. For now, I'd just like to reiterate this point by saying that "one trainer, one animal" is one of my absolute golden rules for rescue and rehabilitation work. For this to be successful, communication between staff regarding where each animal is at in his/her training process is also a key requirement. There can be nothing more frustrating than working with an animal who then is worked with by somebody else inappropriately for that animal's stage of training. Such challenges in communication often cause considerable team conflict, which in turn leads to more inconsistencies and confusion for the animal.

Every single time a horse is handled, they are learning and yet all handling is unlikely to be able to be done by the same trainer. Therefore, for the real world of rescue and rehabilitation, we can consider "training" to be when certain behaviours are actively trained as opposed to, for example, handling such as leading a horse from stable to field if they are generally used to being led. However, this highlights the need for centres to truly invest in behavioural education for all staff handling equines, alongside having very clear organisational policies, guidelines and training procedures regarding handling equines and in dealing with unwanted behaviours. This is where behavioural science can provide a great deal of conformity as it is not subject to marketing of a particular trainer's approach to training. Training staff in the behavioural science of equine learning and behaviour helps to remove the ambiguity of process.

Smaller centres tend to successfully train a smaller number of staff and to overcome differences of opinion within that team. The larger the organisation, especially those with teams across various locations, the harder it becomes for the organisation to ensure a consistent approach. An area of wise investment is therefore staff training, and good management to ensure that the training is being applied. This can then create an organisational culture that firmly has the animal's interests at heart; consequently, such a culture is resilient to changes in staff as the processes become embedded.

Smaller or larger centres can benefit from having either a member of the team who is seen as the leader when it comes to behaviour interactions with the animals or through the employment of a full-time behaviourist who has the skills and authority to inspire other members of staff to use behavioural science. Many organisations make the mistake of employing a "trainer" rather than an appropriately qualified behaviourist. As covered in previous chapters, many methods of horsemanship are not rooted in the science of behaviour or welfare and can ultimately delay or derail the rehabilitation process with very sad consequences. The individual trainers involved may well be skilled, well-meaning, ethical and considerate, however, often they bring a limited set of tools to use with the wide range of behavioural challenges that rescue organisations face. Being in the public eye, rescue organisations need to ensure that their methods of training and handling are ethical and can withstand the robust scrutiny of a growing number of interested parties who can be extremely vocal when it comes to the ethical handling and training of equines.

For the past three years, Maisie Wake has been the main equine behaviourist and massage therapist at Munchkins Miniature Shetland Rescue centre. Maisie describes how they work with the ponies:

Munchkins achieved charity status in February 2015 and is solely run on donations and grants. It is a family-run charity, with a small fleet of volunteers who come to help with the ponies on a regular basis. There are currently approximately 40 resident ponies. For ponies with behavioural issues, I usually work closely with that pony, along with the charity owner and her husband. Once they have a clear understanding of what needs to be done, we discuss which individual volunteer may be able to work successfully with that individual pony, initially alongside me. By pairing up ponies and volunteers, it gives that pony the chance to have consistent and personalised handling. It also gives each of the volunteers the opportunity to build a relationship with that pony, which hopefully increases their commitment to that pony's success and creates motivation for the individual volunteers to do more work like this.

Realistic expectations

Another key element with regards to rehabilitating equines is the importance that everyone involved has realistic expectations about what is possible, how quickly that might happen, how smoothly that process might go and if you'll ever get there at all. If you get really good and the three elements mentioned earlier are put into place then you might see some rapid progress but rapid progress shouldn't really be our aim. We should aim just for progress because once we aim for rapid progress, we are going to put pressure on the animal and

Photo: Jenni Nellist.

on the staff involved in training, which leads to conflict and affects the very process we're trying to improve.

Given what we've already said about the difficulties of creating the correct environment and ensuring that everyone involved in rehabilitation works in the same way, you can see why we should perhaps lower our expectations of what is possible in this work and how long it will take. We should never give up on an equine or presume that they can't be rehabilitated but we must take into account the limiting factors that the rescue centres generally find themselves operating under. When we take the pressure off, a more relaxed atmosphere develops, which in turn often leads to more rapid progress in the animals' rehabilitation as handling becomes focused on the animals' individual experience rather than simply pushing them mechanically through a set of tasks in order to fill in a tick box that can be entered onto the computer.

Being able to treat the animals in their care on an individual basis is one of the challenges that larger equine rescue organisations face. Meeting individual needs for feeding, social grouping and physical requirements is always a challenge when space is at a premium. However, treating equines in their care as individuals in terms of the behavioural interactions between staff and the animals is much more achievable. Individual programmes can be created for each animal. Time can be taken to allow each animal the appropriate consideration during all interactions. Meeting individual needs in this way is realistic and the concept should be embraced at all levels of an organisation.

One more element that is valuable for those involved with rescue and rehabilitation to understand regarding expectations of being in this line of work, is the element of "enjoying the journey". Rehabilitation can often be a very long road for the equines coming into rescue centres. During any animal's rehabilitation, there might well be many setbacks, as well as the consistent challenges of time and environmental pressure, and perhaps organisational pressure to move animals through the process. These daily challenges need to be overcome in order for handlers to be fully present with the animal during training.

When we consider an individual equine with a single owner whose focus is likely to be on the well-being and success of that one animal, and as that person has the autonomy to choose what they think is best for their individual animal, it can seem as though the individual owner has an advantage over rescue organisations. However, when you dig deeper you find many individual owners are confused by which particular training method they should use with their equine or how they should overcome problems and often have a very limited experience of the practical application of the science of behaviour. So, the privately-owned horse might actually receive mixed messages from their handler – the owner is unlikely to have a set of protocols and processes as described above. This must be considered in the rehoming element of rehabilitation – how to safeguard against the recently rehomed horse being subjected to training approaches and inconsistencies that might jeopardise their continued progress in their new home.

Rescue organisations can seize the opportunity to explore up-to-date, science-based approaches and make an informed decision to start using the science of behaviour as a guiding principle. The challenge is getting everybody on board and consistently working to the same principles. When successful, the rescue centre will find themselves at the forefront of the most ethical, innovative welfare and behaviour-focussed thinking in the sector.

This means that people engaged in rescue and rehabilitation work generally have a calm and patient character with a genuine desire to see animals in their care succeed. That pretty much goes without saying, however, in order to succeed, being calm and patient is generally not enough and it's the ability to stick with the process and see the animal through that is a greater determinant of success. In essence, this could be considered "grit" – our ability to keep going and to see things through even when they are difficult. Developing this resilience within teams, to keep on going, is extremely important and here again behavioural science can provide some support in the form of successive approximation or "shaping".

Good days and bad days

In order to provide some insight into the challenges that are faced by staff working in rescue organisations, I'd like to share with you the concept of good days and bad days. This is something that I am often asked about by staff during training sessions; it seems that some days they feel their training goes really well and other days it doesn't go so well but there seems to be a consistent cycle of good and bad days. People working in rescue organisations are keen to understand why these good and bad days occur with such regularity when there appears to be no obvious cause.

What I can share with them is some obvious insights into the likely causes of these fluctuations between good and not so good days. I start by explaining, as we've discussed already in this chapter, that due to the environment in which we are often keeping our rescued equines during their rehabilitation, and when combined with the type of animals that they are generally dealing with, it is pretty normal for equines to experience day-to-day fluctuations in their behaviour. And this is especially true when they're being asked to overcome previous poor experiences, pain, bad handling and humans with a lack of experience. This is just part of the story because what I observe during my work with rescue organisations is the "five and twenty rule" – if you don't stop after five minutes then it's going to take you 20 minutes to get back to where you were after the first five. In terms of behaviour science and training, a behaviourist would recommend short sessions that finish on a good note and make learning easy. My own experience teaches me that we generally get something positive in the first five minutes of the training session, at which point we can either stop the session completely, or take a break and continue with another short session.

More traditional training approaches advocate longer sessions in order to consolidate learning through repetition. Handlers familiar with this type of training are often keen to continue to do more training after the initial five-minute success has occurred. As they continue to train, the animals' comfort zones are stretched, the learning becomes more difficult and generally become more uncomfortable, at which point they often revert to their previous and wanted behaviour in an attempt to solve the problem they currently face – that is, the handler's interaction with them. During this phase, the animal's behaviour deteriorates and the trainer might stop the training session to avoid what appears to be a worsening situation, which reinforces the behaviour further. If, however, the handler continues to work through this burst of unwanted behaviour it might take them up to 20 minutes before they can return to the behaviour level that was displayed during the initial five minutes at the start of the session.

It is easy to see why this happens when the handler has a perceived pressure of getting as much done as possible with the animal in the fastest possible time, in order to help them progress. My perception is that the handler has wasted at least 15 minutes of their valuable time, which could have been spent training another equine. Furthermore, the experience they've given this animal has a large negative element, which despite finishing on a positive note, overall negatively influences the animal's perception of the handler. In this way, the animal is likely to have a less successful training session the next time – in other words, a bad day.

The training element

Chapter 3 provided an introduction to training with behaviour in mind. The elements of behaviour science such as positive and negative reinforcement, punishment, successive approximation, comfort zones, extinction bursts, spontaneous recovery and systematic desensitisation, to name but a few, can be used in all training – there's no difference in the rescue and rehabilitation setting. So, for the sake of this chapter, I'd like to focus on a couple of areas that I think have the greatest potential for improvement in equine handling during rescue and rehabilitation, namely positive reinforcement and successive approximation.

Positive reinforcement: rewards

Positive reinforcement is crucially important in the rehabilitation of equines. It is now widely accepted in the dog world that rehabilitation is much more effective and ethical if you use positive reinforcement – in the equine world, we are slowly beginning to catch up. Positive reinforcement and its successful use is a key factor in retraining equines for two reasons. First, it can have a calming effect on the animal depending on the reinforcer you are using and, second, it shows the animal what to do. If you can tell the animal what to do it gives them a clear experience of what works and the behaviour they can offer again in the future. This positive message helps with the conflict that equines suffer from where they want to show interest in human handlers but due to previous experiences are nervous about what an interaction might lead to. Positive reinforcement creates animals with an "optimistic" attitude – they are more likely to offer behaviour in the chance it will result in a reward. And it creates a positive attitude in the trainer, as the successful steps towards the animal achieving the trainer's aims can be focussed on.

Again, the challenge with aiming to embed the use of positive reinforcement as the basis of training in an organisation is that it is not widely accepted or understood within the equine world and getting staff to embrace this move towards more modern, ethical, innovative training not only goes against the way many people in the equine world have been educated, but also requires the development of personal characteristics and mindsets that are needed in good trainers.

What can I reward with?

You can use food or you can use scratches, even "freedom" to a degree, but all these require the animal to be comfortable in your presence, to take food from your hands without being nervous about it and to enjoy being touched by a human. However, if you are working with extremely nervous and fearful animals that lack experience of human interaction, they are unlikely to be ready to take food from your hand or to see scratches from a human as being positive. What can we do in those situations? Well, we could throw or place food on the floor so that the horse does not have to take it from our hands, but that might introduce conflict – the horse might want the food but not want to be in our vicinity. That approach, therefore, might not be as ethical as it would seem. In terms of the learning definitions introduced in Chapter 3, we are unlikely to be able to use habituation as the animal is already fearful of humans. We can try to use systematic desensitisation and counter-conditioning to gradually approach the animal and introduce them to being handled, but in practice, negative reinforcement is difficult to avoid, as we use trial and error to work out what the horse is comfortable with. If used correctly and in a non-escalating way, negative reinforcement might be essential in communicating with the equine in the early stages of rehabilitation.

Successive approximation (shaping)

It's been my experience that most rescue centres have a very informal plan regarding how they will rehabilitate their animals. In general, that plan covers teaching the animal to be caught, to lead, to have their feet done, to be able to have their bloods taken and to be groomed. Those rescue centres who "produce" riding horses have additional requirements for basic ridden work. Such plans tend to be very general and very dependent upon the member of staff who is undertaking the training. Most centres do not have detailed written training plans, either generically for these areas of handling or individually for animals that have unique requirements. When I train staff at rescue centres, this has been one of the biggest areas we have focused on for the following reasons.

Successive approximation or shaping has been covered in Chapter 3 but is worth revisiting briefly. It is simply breaking down the final, desired behaviour into small, manageable learning blocks or chunks. It's pretty obvious to most people that this is essential training – the difficulty is getting people to realise how small the learning blocks are and when to progress to the next level. It is common to make the steps too big, causing frustration and stress in the animal, which interferes with the learning process. Shaping is completely key in the successful rehabilitation of equines that find themselves in rescue centres.

Let's take a moment to imagine the demeanour and behaviour of an unhandled equine with negative experiences of humans, entering a rescue centre and then imagine the same animal as we expect them to be at the point they are ready for rehoming. It is very easy to recognise the immense amount of work and training that needs to go into all the elements that are required for this animal to live happily and comfortably within the domestic setting. Having to break down the final goal into the small learning blocks can seem like an endless task, but as a minimum might consist of six different training plans each containing between 80 and 120 steps.

There are two elements to shaping: the carefully written shaping plan to connect successive approximations together to reach the final, complicated behaviour; and the individual actions that are reinforced during a training session to increase the likelihood of the behaviour happening again. For me, the ability to understand and practise shaping is vitally important in the rehabilitation training of any equine but particularly of those who have unwanted behaviour or have received harsh, aversive training in the past.

Having a written plan has many advantages

- It provides focus and consistency of training over long timescales.
- It allows us to see how far the animal has progressed in training.
- It enables us to tick each completed step, rewarding us for our efforts and providing motivation.
- It allows gaps in training to occur without the loss of direction or consistency – we can return to the correct place on our shaping plan, even after several weeks' break from training.
- It helps to provide discipline – we are less inclined to skip written steps or to rush the training.
- It allows us to deal with problems by moving back down the steps to an achievable level before proceeding again.
- It ensures we do not miss out any vital steps.

In the rescue centre context, having a written plan has the additional benefit of being able to share the plan with the team; this helps to keep everybody aware of the animal's progress and of what should not be attempted with the animal based on the training not yet received. It also serves to increase health and safety by having a record of each animal's training and managing people's expectations of the animal's performance. This should also minimise the risk of asking too much too soon, which could be potentially dangerous.

I am so passionate about the importance of creating written shaping plans that I have developed a series of ready-made plans that can be worked through. More information on the construction of a written shaping plan can be found at www.hartshorsemanship.com/books.

The best rescue centres are the ones that have clear, consistent, written shaping plans for those animals undergoing rehabilitation; those plans are available for everybody to see and are regularly updated and used enthusiastically by staff that have shaping as the very core of their training routine. Each and every behaviour that is required to be trained requires an individual shaping plan, but these do not have to be used in a linear fashion and a competent trainer could be working on several shaping plans at once. It is important to know that each session the handler has with an animal, they should only work on one shaping plan and if choosing to work on several shaping plans, there should be small, short breaks between the different plans to minimise confusion and optimise success.

Some of the written shaping plans required for general training

- Standing still – always teach this first
- Being touched
- Yielding to pressure
- Leading
- Catching
- Moving off pressure
- Touching body all over, sensitive areas
- Vet prep
- Feet

Using the shaping plans gives clear focus and direction during the animal's progression through their rehabilitation. A written plan provides a platform for consistent handling and allows for the natural problems that arise in busy rescue organisations where animals aren't always able to be worked with every single day. The plans can be tailormade for individual animals or more generic for animals with less unwanted behaviour. Whatever the requirement of the animals, shaping plans really are a broad template that could be applied to a remedial animal or one with little or no experience.

Other ways rescue centres can apply behaviour in mind

As well as applying the key elements for successful rehabilitation described so far, there are many other parts of the rescue process where rescue centres can operate with behaviour in mind. Examples include: managing a frequently changing herd; rehoming with behaviour in mind; outreach work to prevent horses coming into the centre in the first place; post-homing support; campaigning and outreach messaging to address the wider causes of problems for which the rescue centre is needed. In this section we will explore managing a changing herd and rehoming.

Managing a changing herd

Whether in reference to preserving the natural needs of the animals, or for safety, or as a basis for allowing horses to be in the best possible "mindset" for training or in reference to minimising the likelihood of injury, allowing horses to form bonds with each other and minimising disruption to those social attachments has been mentioned many times. Due to the continual arrival of new equines into their care, and equines leaving as they are rehomed, rescue organisations face the challenge of matching new equines to existing groups within existing management systems and managing the safe introduction of each new animal into already established groups.

Horses are often kept in individual stables or stalls and prevented from social contact with other equines for long periods of time. As discussed throughout this book, meeting the social and environmental needs of each animal is extremely important. Keeping horses in individual stables limits their behavioural repertoire and impacts on their ability to interact socially and, as such, provides reduced opportunities for expression of natural behaviour. There is also a physical difficulty in managing a large number of equines in individual stables as this requires a very labour-intensive system and is expensive for rescue organisations to maintain. Consequently, many rescue organisations provide group housing for the equines in their care, which, in principle, enables equines to be kept in a more natural setting and allows organisations to meet more of the animals' behavioural needs.

Most organisations manage their herds by creating groups of elderly animals, groups for young animals and groups for those with specific behavioural needs. However, if we use the behaviour in mind concepts, we need to explore the individual's needs a little more deeply by considering their temperament, levels of confidence, natural tendencies towards leadership or following other members of the herd – in essence, their individual distinctiveness (personality) needs to be considered.

Introducing a new equine into a group is not only stressful for the individual but also for the entire group. This is exacerbated in the domesticated setting of the rescue environment because resources such as food, water, shelter, companionship and space are likely to be limited and can all be conflict points between the groups. In an ideal world with huge fields, large barns and sheltered areas where each group had access to a variety of different areas to rest and explore, there would be less concern regarding resource-guarding behaviours. Sadly, for many rescue organisations, this utopia is far from reality due to the physical cost involved and the massive requirements for space. In this respect, the individual rescuer with a small number of equines may be better placed to provide this ideal environment.

We tend to view the aggressive behaviour around the introduction of a new arrival as part of the normal part of horses' communication, and in all settings it is common to simply turn a horse out into a field with the other horses and "let them sort it out". However, if we consider the behaviour of feral horses, we quickly appreciate that such sudden introduction is highly unlikely to happen outside the domestic setting. In the wild, horses home range and at points the home ranges of different herds might overlap. However, the horses wouldn't suddenly come across each other; their movement in their home range would be such that the first indicator that horses from another herd are in the vicinity would be signs such as coming across droppings, areas where they have rolled and seeing them from a distance. The herds are not suddenly confronted with new horses and have a much greater area to move away from potential conflict situations. Furthermore, within feral herds, aggression is not commonly observed. Joel Berger, who studied wild horses in the USA states "On average less than one individual per three hours of observation would be involved in food related aggression even during the most stressful season" (Berger, J. 1986). This is generally not the case in our domesticated horses. Therefore, considering that the rescue environment necessitates rapid introduction of horses, and that domestic horses are more prone to showing "aggressive" behaviours due to limited resources, we must carefully manage the introductions between horses.

Managing the introduction

The only way to minimise stress and the risk of injury during introductions, is through a process of gradual introduction of horses to new environments or herds. This should take place over at least a couple of weeks and more if needed. Some horses are introduced rapidly to new herds and all is well, but this is always a risk, especially when animals with a history of perhaps neglect, starvation and poor socialisation to other horses are involved. In my opinion, the risk of "hoping for the best" is too great to consider.

Horses that are to be introduced to a new herd should be first allowed access to each other over a safe barrier, such as a field fence or gate, where the barrier is electric fence, and if there are any concerns that the horses may fight or strike out, a double width of electric fencing placed at least 1.5 m apart will prevent the horses from being able to reach each other.

As horses rely on smell for part of their

communication, transference of faeces is a valuable part of a behaviourally-minded introduction process. Faeces from the new horse can be placed in the area with the rest of the herd for them to explore, and vice versa. It is understood that horses can determine the sex, approximate age, level of stress and other information through such olfactory contact. Depending on the animals' behaviour, and plenty of observation is the key here, if all is progressing well the dividing fences can be moved closer together allowing gradual contact over the course of a week. During this time, taking the horses for walks, in hand, together with competent handlers and ensuring sufficient space between them while walking will also be a very useful introductory exercise.

If the animals are comfortable with each over the fence, erect an electric fence paddock inside the existing field, which allows the new horse to be safely "in" the herd but allows a safety space if there is a problem. Place the new horse in the small paddock and then allow the rest of the herd access to the field; this should be under strict human supervision and preferably one horse at a time.

The observer should be looking for signs of communication and submissive behaviours. Too often, people focus on considering behaviours in terms of a loosely defined concept of "dominance" when what is more important in this situation is submissive behaviour. Submissive behaviour actually protects the herd and minimises conflict and is extremely important in herd life and social acceptance. You should be seeing horses greeting each other over the fence without threatening faces or gestures, animals should be happy to graze in each other's presence and there should be no excessive conflict over food. You want to see horses communicating calmly and with small signals, moving away from each other and reacting appropriately to each other's requests for space.

Depending on the observations, you may then move to a process of placing the new horse in the field and gradually adding one horse from the herd at a time, usually starting with a well socially-adjusted horse with good communication signals. Allowing the pair of horses to spend some time with each other before introducing the new "pair" to successively more horses in the herd is advised. Always ensure that the new horse to be introduced has had individual access to the paddock/field environment before introducing to the herds; this allows the new horse to check out the environment and to observe any hazards or obstacles that might need to be avoided later on. Have the new horse in the field and add the others, rather than adding the new horses to the existing herd.

We can never be sure that horses will get on completely and for that reason, once introduced to the herd, observe the behaviour during the course of the day before deciding if it is safe to leave them together overnight. If you sense there is likely to be any aggressive behaviour, then best to remove the new horse from the environment and repeat the process through the daylight hours the next day.

Managing introductions in the rescue situation should be something that every organisation takes extremely seriously; policy and procedures should be in place to set best practice to minimise the risk of injury to the animals and staff, but also to reduce the levels of stress and anxiety that are placed on the animals when they have little free choice in their selection of band mates.

Maisie, the behaviourist who works with Munchkins who we met earlier, describes how they introduce new ponies:

> On arrival to the charity, all ponies must go through a two-week quarantine. They are

Photo: Suzanne Rogers.

then slowly introduced to select ponies, before joining the main large herd. This is done as incrementally as possible, with introductions done at a slowly decreasing distance, with a barrier between the ponies. After I made the charity aware of the importance of the olfactory system, and the value of their getting to know each other's scent through faeces as well, they came up with a solution that works for them. Sometimes a "poo transfer" can be done, swapping the new pony's and resident pony's poo. Where this is difficult at Munchkins, with a larger number of ponies to work with, instead they allow the new ponies to explore their potential herd mates' vacated field, and the resident ponies go into the newcomer's field, so that they can familiarise themselves with each other's scents before meeting.

Any forage put out into the field is always spread over the entire field, and the handlers get quite imaginative with this. They also never feed from buckets in the field, ensuring that the handful who do need specific feeds are brought in and fed individually. Everything is aimed at an atmosphere where there is little reason for resource guarding. Mares, geldings, young and old live together, unless this poses a difficulty within the herd. A separate small herd of "oldies" is now slowly developing, for those who are too old to manage within the herd but get along better in a smaller and quieter group. There is also a separate group closer to the owner's main farmhouse, which is set up perfectly for managing laminitis and metabolic issues.

While it is important that ponies are adopted out, a portion of the main herd will always remain at Munchkins. This gives the herd an element of stability, which provides a calm and stable environment for the new ponies to arrive into. Conversely, the occasional removal of ponies to go into a new home is always done sympathetically and strategically, aiming at avoiding upheaval to either those being moved, or to the main herd.

Rehoming

The goal of most rescue organisations is obviously to rehome as many animals as possible. Rehoming reduces the financial burden to the rescue organisation while creating space to take in future cases that require support. Rehoming is only a positive outcome for the animal if the home that is selected is suitable and can provide the correct care for that individual animal. It is only positive for the new owners if the animal is appropriately prepared to a level of handling that will allow them to safely interact and succeed in the new home.

As discussed earlier in this chapter, the pressure to get animals out onto a rehoming scheme can mean that the process of preparation is rushed. This is most often seen in the lack of generalisation that is included in the training process. In general, the equestrian world makes the mistake of presuming that an animal who has been exposed to one situation, is then prepared for the same situation wherever it may occur. For example, the animal who is learning to walk over a tarpaulin in an arena is mistakenly considered capable of doing so in any other environment where they may meet a tarpaulin in future. Once we understand the need for generalisation in the handling and preparation of equines for rehoming it is easy, but more time-consuming, to add additional experiences in a variety of locations that prepare the animal more fully for their new home. To continue with our example of the tarpaulin, once the animal has walked over the tarpaulin in the arena, the tarpaulin must then be encountered in different positions within the arena, then outside the arena, on the way to the stable, in the field, on the lane or track and possibly in the stable yard. This way, the animal learns that wherever they meet the tarpaulin in the future, the behaviour of the tarpaulin is likely to be the same and confidence is increased. When repeated with many other obstacles and objects, the horse builds up even more generalised confidence with respect to encountering new obstacles anywhere they might be. A wide variety of experiences must also be generalised using a similar process – having feet handled in the stable, field and yard, by different people, and so on to truly be confident that the horse is set up for success in a new environment.

This work prepares the animal more fully for the arrival in a new home with new handlers who are likely to be less aware of the behaviour or requirements of the animal in terms of training. Hopefully, the equestrian world will continue to become more aware of the use of the science of behaviour in training and handling equines but for the time being, we are in a situation where animals need to be prepared by responsible equine charities more fully to allow them to succeed in the home. There is of course the additional concern of newly rehomed animals injuring their new owners and the equine charity being sued as a result, and this should be incentive enough to ensure that a generalisation programme is part of every equine rescue organisation's rehabilitation process. This also ties in nicely with the concepts of shaping and having written shaping plans, as it is easy to have standard practices of generalisation to obstacles and a variety of different situations.

Preparation for the new home obviously depends on the organisation's ability to understand the individual requirements and situation of that home. The more that can be done to prepare the individual animal for what they are likely to encounter in their new environment, the better. For example, a horse moving from a group housing situation in the rescue centre to being stabled individually in their new home would require preparation to ensure that the horse copes with such a change (ideally of course the benefits of group housing would be understood by the new home!).

Other novelties that the animal may encounter in their new home such as small stock, children or pigs can be potentially included within the preparation stage before the animal reaches the new home and this is where the generalisation element of training can become especially important. While it is appreciated that not every rescue centre can keep pigs specifically for the preparation of rehoming animals, a little imagination and the use of models, neighbours or visiting opportunities can create realistic alternatives.

Holly Roberts, a groom at RSPCA rehoming centre Lockwood in Surrey, UK explains her role in preparing horses for life in their new homes:

> We do our best to ensure that each horse/pony/donkey has as much confidence as possible with the everyday handling we would expect them to receive in their new homes (including grooming, feet handling, leading, farrier and vet visits). We aim to give them the building blocks to learn how to cope in new and potentially worrying situations once they get to their new homes. We try to do this by engaging them in stimulating and fun activities such as going for walks or exploring our obstacle courses. We keep our handling sessions short and sweet in the initial stages, quite often just being with the horses and allowing them to come and explore near us and allow us to touch them with no expectations at all. We can't prepare them for every situation but by using positive reinforcement and a lot of time and patience, our horses leave Lockwood with a bank full of positive experiences that we hope they will draw upon once they face new situations in their new homes.

Matching people and equines

From personal experience, the situation when previously well-handled and suitable animals are returned from homes due to behaviour problems, is a great source of frustration to the centre staff that have carefully and lovingly prepared animals for their new home. As well as preparing equines for their new environment, it is also important to prepare people for the new equine. It is very easy to assume that potential owners have the knowledge required to take on a horse. Given that there are more horses needing homes than homes available, it is understandable that organisations can overlook potentially serious differences or personality clashes in the desire to rehome an animal.

As a gold standard, some rescue organisations insist that new owners attend a training course before any animal will be released to them. Such training would ideally include general information regarding management and training but also personalised training for the owner with the specific animal that they hope to rehome. There is often reluctance by organisations to run such formal courses due to the fear that such requirements might reduce the number of potential new owners. It is more typical that organisations rely on a shorter personal introduction to the animal, possibly with a trial handling or riding session and a home visit, all of which are to be encouraged, but my personal feeling is that more in-depth training is of great benefit and reduces the risk of the animal being returned.

Holly Roberts describes how their animals are matched with new owners at Lockwood:

> After adopters have filled out an application form and we have spoken to them on the phone, we do our best to match horses and owners considering their needs, experience and expectations. When they come to view the horse that we believe will be the best match, they spend time grooming, leading and generally just being around the horse before deciding whether to reserve the horse pending a home check. The grooms will always have an honest discussion about the horse's behaviour and the expectations the adopter should have for the horse. With horses that aren't quite so "straightforward", we will ask adopters to come back and visit at least once more to see the horse being brought in from the field and how he/she interacts and behaves with the grooms and other horses from

the moment the head collar is put on to be brought onto the yard until the horse is returned to the field. This gives adopters a real insight and can often be a much more realistic viewing than having a grooming session on the yard with the horse.

Holly continues:

The right environment for our horse is so important and we will have already made a decision about the lifestyle requirements the horse has with whoever adopts them. We have categories regarding what lives our horses are suited for: a companion only due to age/injury or behaviour problems that we feel would be worsened by the expectation of ridden work; companion for now with the possibility of ridden work should the horse or pony become more confident and show no signs of discomfort (physical or mental) during the backing process; and horses that we feel are physically and mentally prepared for a lifestyle that includes ridden work. We find that our adopters choose to rescue for a reason and are usually very amenable to the lifestyle that the horse needs to have and tend to fall in love with them for their gorgeous personalities rather than how they will be "used".

Maisie describes the rehoming outcome of two of Munchkins' ponies who had specific requirements:

We are careful that nervous ponies, or those with specific behavioural issues go to experienced homes, or those which are happy to continue with behavioural training. Two ponies, Pebbles and Bam Bam, have gone out to a local home, where part of the arrangement was that they would continue to have both behavioural and bodywork sessions from me. Pebbles and Bam Bam were set to be permanent residents at Munchkins, because Bam Bam has asymmetrical hips, with consequent stifle issues as well, and Pebbles has juvenile arthritis. This was a shame, however, because Pebbles was extremely people-friendly and potentially one of the few ponies who would be good for a family to rehome. Bam Bam is a tiny, very physically appealing pony, who would easily find a home. This has always concerned me, because he is a child magnet, but is actually rather nervous. For this reason, I strongly emphasised that if the novice family in question were to take them on, it would be good for them to have regular training sessions with me, as well as bodywork sessions for the ponies too. This would also show a commendable level of commitment which would demonstrate a good level of responsibility towards the ponies' needs. This has been a huge success. The children are very exuberant, and without immediate guidance, this new home could have been a disaster for a pony such as Bam Bam. Instead, Pebbles and Bam Bam now flourish in their new home, benefitting from more individual care and attention than they would get within the herd. I should add that training sessions with this family began at Munchkins, before the ponies were rehomed. This is a nice way for them to become familiar with the family before having to move to an unfamiliar environment.

One of the most important aspects of preparing the new home for the arrival of the equine is to manage the expectations of the new owner. People are enthusiastic and keen to do the right thing but often underestimate the length of time it will take for the new equine to settle in and, as a consequence of this, do too much with their new equine partner in a short space of time, which has the potential of damaging the relationship at best and at worst causing serious injury.

Creating an eight-week settling in plan with the new owners can be extremely beneficial. During this initial period, it's important that the new home focuses on (ideally with a shaping plan) building trust and developing the relationship between equine and human rather than trying to develop skills and experiences such as riding or hacking out. In this period, the training in generalisation of specific handling and experiences can be continued, building on what was done in the rescue centre.

From personal observation and feedback from owners of rescue equines, it normally takes around a whole year before owners report they feel confident and comfortable in their relationship with their adopted animal. The massive changes that the equine faces in entering a new home and the length of time it takes them to settle in to the new routines and environment cannot be overstated. Personally, I feel it is the rescue organisation's responsibility to apply some of the key principles of the science of human behaviour change in helping people to have realistic expectations and in understanding how to meet their animal's needs. This will be covered in the final chapter of this book.

Making it clear that the transition from rescued to rehomed is part of the animal's journey and requires continuing work on the part of the owner, is vital. As part of this continuing journey, it is important for the rescue organisation to offer behavioural support to the new owners whenever it is needed and this should be in the form of a qualified and experienced behaviourist who can assess not only the animal's behaviour but also that of the human, and who will work to find a solution that is individual to the situation rather than just applying a blanket approach. Problems will occur in the new homes and working with behaviour in mind ensures that these problems can be tackled early before they become ingrained.

Pocket is rehomed

At the beginning of this chapter, we met Pocket the pony. Pocket was lucky enough to find herself in a rescue centre that had behaviour at the heart of everything they did. The rescue staff found that when she was kept in an environment that met her needs, with people who had a compassionate, knowledgeable approach, her behavioural issues faded with very little active training. She was carefully integrated into a herd and formed a close relationship with a pony called Brambles. The grooms created an online profile for Pocket that reflected her nature and struggles and it was spotted by a potential new owner, a dog walker called Ingrid. After meeting Pocket and learning extensively about what horses need from the rescue centre, Ingrid ended up adopting Pocket and Brambles so that they could continue to be together in their new home.

Final thoughts

Rehabilitating and rehoming an equine is a tremendous experience, if you want to enjoy the journey and you have the grit to complete it. In this chapter, I introduced what I consider to be the three vital elements for successful rehabilitation: the right environment to ensure a relaxed, calm animal that you can then begin to work with; a relaxed, calm human being who is not rushing and is consistent in behaviour throughout the training experience; and the ability to train behaviour correctly through the use of applied scientific knowledge. I described the value of written training plans to ensure the steps you take are small enough, in the right order and allow for the animal to develop their skills. That is the magic bullet – there is no short cut to rehabilitation. We can't change human behaviour easily and quickly and if we can't do it in humans, supposedly the most intelligent species on the planet, why would we expect to do it quickly with horses and donkeys? What my experience does tell me is, if we get these things right, the ability for animals to rehabilitate themselves is phenomenal.

Rehoming organisations face many challenges in the rescue and rehabilitation of equines and as we've seen in this chapter, the nature of their work provides advantages and disadvantages to the behaviour of the animals in their care. It is clear to see that with a commitment to behaviour, any rescue organisation can vastly improve the way in which it keeps, trains and rehomes animals. The challenge is for organisations to accept that working with behaviour in mind requires a consistent approach from all staff and a dedicated commitment from the organisation itself to remove time pressures and focus on quality of rehoming rather than quantity, while recognising the value of preparing equines for homes.

Reference

Berger, J. (1986) *Wild Horses of the Great Basin.* Chicago University Press, Chicago, USA. p.156.

Chapter 8
Vets
Amber Batson

Behaviour is affected by diseases including those causing pain, and certain behaviours can lead to the development of disease. This chapter explores the link between behaviour, disease, pain and stress to illustrate the importance of having behaviour in mind for veterinary professionals. It also includes ways that owners can prepare their animals for procedures that vets might need to undertake.

Hunter's story

Consider "Hunter" a 15-year-old Irish Draft cross gelding who lives in a livery yard in a leafy English village. Throughout the autumn and winter of every year, Hunter lives in his stable for 24 hours a day because the local ground is clay and once the UK's wetter seasons begin, the ground becomes seriously muddy and sticky, and turning the horses out at this time risks injury and significantly damages the land. Hunter has a long-standing stereotypy of weaving. In the last six months, Hunter has started to become aggressive when having his stable rug put on and when having his front feet picked out. The physiotherapist recently assessed Hunter and found his neck to be very stiff and so the vet was called and radiographs (X-rays) of the neck were taken, which have revealed arthritis. The repetitive action of weaving over multiple years is likely to have been a major contributing factor to the development of neck arthritis in this case. By understanding the causes of behaviour problems and the ways disease affects behaviour, we should be able to help Hunter with his recent development of aggressive behaviour.

There are multiple ways in which disease processes in the body can contribute or even be the sole cause of behaviour problems. The most common "disease" contributing to behavioural problems in the horse or pony is pain. Pain itself clearly isn't a disease: pain is a symptom of an underlying disease. Common examples of reasons for pain in the equine are laid out in Table 8.1 (it should be noted this table is for illustration only and not intended as an exhaustive list).

The link between pain and the development of behaviour problems

There are several ways that pain and behaviour are linked due to changes in the body that occur when a tissue is injured and inflammatory chemicals are being produced. Think of yourself – if you have ever hurt a part of your body such as your back or your neck, you don't want others to touch you, or even approach

Table 8.1: Common examples of reasons for pain in the equine.

Acute	Chronic
Gut pain (colic)	Stomach pain caused by ulceration
Joint pain – strains	Arthritis
Muscle pain – strains – other myopathies such as "tying up" or atypical or seasonal myopathy from toxin ingestion	Chronic muscle/ligament pain
Acute uveitis (inflammation of the middle layer of the eye)	Chronic uveitis
Mouth ulcers or dental root pain	Chronic dental problems such as pulp exposure (see Figure 8.2 for a picture of a mouth ulcer caused by dental disease)
Laminitis	Chronic laminitis with associated pedal bone position abnormalities
Sinusitis – such as from infection or trauma	Chronic sinusitis from prolonged infection
Nerve pain from inflammation or infection	Chronic nerve pain from long-term inflammation or infection
Hepatitis (liver inflammation) as the result of acute infection, fats or toxins	Chronic hepatitis associated with long-standing liver inflammation
Cystitis (bladder inflammation) as the result of infection or bladder stone development	Chronic cystitis associated with long-standing bladder inflammation

you, as you don't want them to worsen your pain.

One of the first behaviours we often see in the painful animal is increased self-defence. There are four main behaviour categories we use when trying to defend ourselves:

1. *Flight*: We try to move or even run away from the potential threat.
2. *Fiddle about*: We use language to communicate "Please don't come close, I don't mean any harm to you, please don't harm me". Equines clearly cannot speak to us through verbal language to tell us of their intentions; however, they can use their body language. We call those body movements aimed at defusing a potential threat "appeasements".
3. *Freeze*: We go very still in the hope that the threat will move away.
4. *Fight*: The use of aggression with the intention of intimidating the threat to leave us alone.

When horses have pain somewhere in their body, we are more likely to see one or more of these self-defence behaviours than we usually see in that individual. Recently, there have been increasing efforts to try to describe behaviours and facial expressions that are common features in the painful equine. Indeed, a 2016 project to identify the key welfare challenges in the UK found that "non recognition of stress and pain in the horse" is one of the top four (overview provided in *"Horses in our Hands"*, Horseman and colleagues, 2016). Mullard and colleagues (2017) published the facial expressions of pain describing postures and expressions related to head carriage, nostril and mouth tension as well as ear and eye position. Many of these features are commonly observed as part of the expressions seen when an equine is fearful too, which makes the recognition of pain complicated. The facial grimace as described by Dalla Costa *et al.*, (2014) seems fairly consistent with an equine in pain but not all horses in pain demonstrate their pain in facial expressions. As we can appreciate ourselves, the sensation of pain in the body can vary in how it feels (such as burning, tearing or stabbing, for example) as well as in intensity (low level, moderate level or high level). As a prey species, the equine is extremely good at hiding its pain, after all you do not want to demonstrate potential weakness to any possible predators. Unfortunately for the domestic horse, this tendency to "cover up" discomforts and pains, can affect their welfare, as a fairly high proportion of the population can be experiencing pain and yet the owners, handlers and riders might be relatively unaware.

Dr Sue Dyson (Dyson *et al.*, 2018) estimates that more than 47 per cent of the sports horse population in normal exercise might be lame without any level of recognition from the owner, rider or trainer. Longer-term pain, often referred to as chronic pain, normally meaning that the pain has been present for several weeks or longer, is associated with the production of long-term stress chemicals such as cortisol. We will be looking at the effects of stress chemicals on behaviour shortly, but in brief, chronic stress caused by chronic pain creates a frequently maladaptive cycle in which the perception of the pain worsens, creating a range of side-effects. Such side-effects include the experience of "wind up" – in this situation the stress chemicals switch on, or further "sensitise" the tissues, receptors and nerves responsible for detecting and messaging potential tissue damage to the brain. This means that any potential pain in the body is detected even faster and is experienced at a greater intensity than if pain was not already present. This is true of both the original site of the pain but also in subsequent parts of the body where the individual has pain, even if unrelated to the original area.

Chronic pain sufferers often also demonstrate allodynia – feeling pain in sensations that would not normally cause pain, such as being touched gently on their skin (see the case described in Figure 8.1). Our Irish Draft horse "Hunter" might well be experiencing allodynia when we attempt to put on his rug; touching parts of his body that previously did not cause any discomfort, now results in his experiencing pain, which might be why he can no longer tolerate a rug being put on or adjusted.

Hunter might also demonstrate "hyperaesthesia", another side-effect of chronic pain. In hyperaesthesia, stimuli or experiences that normally only cause mild discomfort, such as having a thin needle placed into the neck muscle for a vaccination injection, suddenly are interpreted by the brain as a far more painful event. Yet, sadly, horses showing fear reactions in the presence of previously neutral situations might be anthropomorphically described as "naughty" or "difficult" as the tendency of a prey animal to hide their true underlying "weakness" can be a challenge to understand.

Animals experiencing ongoing or regular pain will often try hard to avoid the situations that trigger the pain. When ridden, Hunter is likely to avoid turning his head and neck to the side when pressure is applied to the bit with the rider's intention of making him flex his upper body laterally. This resistance to flexion might be perceived as a resistance to the rider's aids rather than the avoidance of painful movement that it really is. It is possible that Hunter might begin to refuse jumps as the neck positions required to balance himself on take-off, in the air and on landing may be painful. Performance reduction or the development of evasive behaviours such as "running out" or bucking on approach to fences can be indicators of underlying pain.

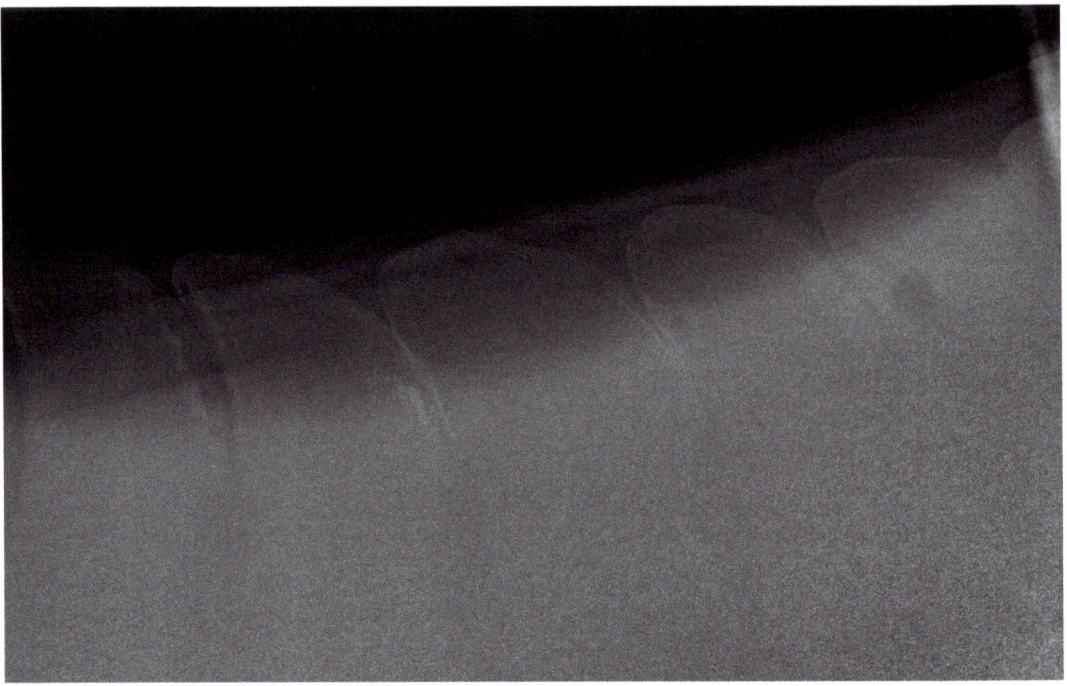

Figure 8.1: The figure shows a radiograph of a horse's back with the condition of "kissing spines" (overriding dorsal spinous processes), which are a fairly common cause of back pain in the horse. This horse was presented to the author with a sudden onset of aggression when being groomed anywhere on his body, likely the result of allodynia occurring alongside moderate focal pain at the site of the overriding bones. Photo: Amber Batson.

Interview with Maruska Aylward-Green MSc BSc MCSP HCPC ACPAT A – an ACPAT Chartered Physiotherapist working at Bridgefield Physiotherapy Ltd, Surrey, UK.

Q. *As a chartered animal physiotherapist assessing and treating equines on a daily basis, how common do you think it is that pain or discomfort go unrecognised in horses and ponies in the UK?*

Very common; I would say that in approximately 70 per cent of our practice caseload, owners unfortunately fail to recognise pain behaviours. The problem is that many clients are unaware of how pain behaviours can manifest and be demonstrated, after all, being a prey animal, the horse is very good at keeping "quiet" about pain, which makes it difficult for many owners and riders. I regularly assess horses and ponies and observe and point out what I believe to be a pain behaviour and am often met with the answer "Oh, but he's always done that!", for example the horse that has always been described as "girthy" to tack up. Just because that is the case, it doesn't mean it isn't pain-related and should therefore be acceptable.

Q. *How many of the equines you are asked to assess/treat are presenting with a behavioural problem which is the main reason for the owner/rider asking you to examine the animal?*

Approximately a third of new patients referred to us either via vet referral or via owner referral under veterinary consent I would say have a behavioural problem. These problems most commonly would include bucking under saddle, rearing, kicking out when asked to canter and not going forwards in the school. These are the most common. Another third of cases are usually performance-related problems, which when assessed we often find are pain-related. The most common complaints here, for example, would be: disuniting in canter, stiffness on one rein, underperforming in some gymnastic ability such as jumping, or struggling to work long and low. These may all be caused by a lameness issue related to pain or pelvic or neck pain for example. The last third of clients book horses in for a general physiotherapy assessment in order to ensure they are comfortable; there is no specific complaint.

As a physiotherapy business we work hard with our clients to educate them on pain signs and I feel our clients are definitely improving in recognising pain behaviours and acting upon them sooner than when I started practising nine years ago.

Q. *When treating the equine for musculoskeletal problems, do you find it is important that the animal is as relaxed as possible? Would high muscle tension due to fear during certain training activities have an impact on the desired effects of any stretching or similar exercises used in an animal's rehabilitation programme?*

It is vital that the horse is relaxed. It is more difficult for the physiotherapist to palpate into deeper tissues if the horse is stressed as well as it being more difficult to palpate deeper structures in horses with chronic or acute pain. In this case, the "hands on" treatment effect will not be as effective. High muscle tension due to fear could also have a negative impact on stretching exercises. I find stressed, fearful horses find stretching with their head between their forelimbs difficult to execute. This then means they do not stretch their back muscles under the saddle area sufficiently enough to mobilise their back up into flexion. These horses also find working with their back on a stretch difficult as in order for this to happen they must work with their heads low and generally, if they are stressed, they resent adopting this posture.

Chronic pain has also been described as the cause of behaviours such as self-mutilation syndrome (McDonnell, 2008). In a review of self-mutilation syndrome in stallions, McDonnell describes a list of physical discomforts causing self-mutilation. In these cases, she describes how treatment of those underlying conditions resolves the self-trauma with no further behaviour modification required.

Diseases that cause or contribute to behaviour problems

Diseases other than pain have been recognised as causing or contributing to behavioural problems in a variety of domestic animal species. It is outside the scope of this chapter to consider all the potential diseases that might affect behavior, but potential examples for the horse would include the following:

- Cushing's disease (also known as pituitary pars intermedia dysfunction or PPID) is a disorder of the pituitary gland located in the brain, in which the gland is not producing enough dopamine, a key brain chemical, resulting in an "unchecked" production of hormones being released by the gland into the horse's bloodstream. One of these hormones is adrenocorticotropic hormone (ACTH), which is the hormone responsible for switching on the adrenal gland to produce the body's main steroid hormone, cortisol. An overproduction of these hormones results in many changes, the most common of which physically is the development of laminitis. In the author's experience (and also mentioned by www.thelaminitissite.org/ppid-faq/temperament-changes-with-ppidpergolide), these chemicals can trigger behavioural change including increased fears, increased startle behaviours and even aggression. Treatment for the condition is most frequently to administer a medication (pergolide), which increases dopamine activity in the pituitary gland allowing for regulation of hormone release from the gland. Successful treatment of the disease normally results in resolution of the behaviour changes.

- Granulosa cell tumours are a form of benign cancer in the female horse affecting the ovary. The tumour causes the excessive production of testosterone and/or some other reproductive hormones and in approximately 50 per cent of cases this results in aggressive, sometimes "stallion-like" behaviour in the previously normally behaving mare (McCue et al., 2006).
- Retained testicular tissue or undescended testicles (rig). This is not a disease per se as the presence of testicular tissue in the male equine is clearly a normal occurrence. However, the development of "stallion-like" behaviours such as increased vocalisation, striking out with the foreleg, mounting and thrusting behaviours and increased tendencies to aggression, are often reported in geldings that have not had all of their testosterone-producing tissue removed during the castration procedure.
- Petit mal or absent seizure epilepsy has been described as causing behavioural change in some horses. Changes include depression, hyperexcitability, irritability and the development of marked "compulsive-like" behaviours, such as manic biting at the ground or compulsive licking/chewing, kicking out or violent headshaking. In dogs, a variety of behavioural abnormalities are seen in those suffering from "absent" seizures – these are changes in electrical activity in the brain that doesn't result in a loss of consciousness, more that during the seizure the animal appears to be "unresponsive" to stimuli in the environment whilst unusual behaviours for that dog, such as aggression, hiding, pacing, whining, and so on are seen around the time of the seizure.
- Brain tumours are not uncommonly cited by people as a cause of more extreme or bizarre behavioural problems in animals including the horse. Interestingly, tumours affecting the brain tissue in horses are extremely rare in comparison to those reported in dogs and cats. Although sudden changes in behavior, such as marked depression or marked aggression, could potentially be related to growths within the central nervous system, research suggests the likelihood is extremely low (Pirie et al., 1998).

Dysfunction of the thyroid gland is recognised in dogs (predominantly hypothyroidism, an underactive thyroid gland) and in cats (predominantly hyperthyroidism, an overactive thyroid gland) with one of the presenting symptoms in both species being behavioural change. Hyperexcitability, marked docility/lethargy and aggression are behavioural symptoms reported in dogs, cats and horses with abnormal thyroid hormone levels, however true thyroid disease in the horse is rare in comparison with its presence in dogs and cats (Breuhaus, 2011).

In summary, whenever a behaviour problem is recognised in any animal, it is important to consider the potential for underlying disease as a contributing factor. In equines, there is currently less attention paid to this than in species such as the dog, yet as the above examples suggest, a thorough consideration of such possibilities may be the most effective way to begin, if not fully resolve, the change in behaviour in many cases.

The link between stress and the development of behavioural problems

Stress is a normal physiological process to help an animal adapt to changes in its immediate environment. Stress can be acute – occurring within seconds of a potentially exciting or threatening stimulus that requires action from the body. Acute stress is accompanied by internal release of chemicals including noradrenaline and adrenaline, which help prepare the body for faster activity as well as an increase in sympathetic nervous system activity: the flight or fight system. These two internal mechanisms increase the individual's speed of reactivity, increase blood flow to muscles to help them move the body faster and increase heart rate, blood pressure and breathing rate as well as decreasing pain perception.

The body is unable to sustain such intense changes for more than several minutes, so the mammalian body developed a second system – the chronic stress pathway, which starts to take over the maintenance of increased activity for a longer period of time. Think about it like this: the zebra recognises that a group of lions is moving around her herd in a "suspicious" way; in less than seconds the zebra has tensed her muscles, lifted her head and neck, started to breath faster and is now turning on her hooves and fleeing away from the lions, along with the rest of her herd. This is acute stress in action.

However, as they gallop across the plain, it becomes apparent that this pride of lions means business; after a few minutes, the zebra notices the lions are not slowing down, the chase is continuing. The lions are persistent. Minutes come and go and the chase goes on. The zebra's heart is beating so fast, her blood pressure is extremely high, lactic acid has built up in her muscles. She needs a new strategy; she needs a slightly longer-term plan. This is chronic stress. By now her hypothalamus and pituitary gland deep in her brain tissue are releasing chemicals to switch on the adrenal cortex (the outside part of the gland) to release glucocorticoids – the main one of which is cortisol. Cortisol has several functions in the body but, in essence, it provides a way for the zebra to find alternative energy sources in her body to keep going. This includes switching off "energy-rich projects" such as building new white blood cells, or producing and maintaining a healthy reproductive tract in order to get pregnant, as well as projects such as growing new thick and healthy hairs. Cortisol also communicates to the brain – don't feel sleepy right now, sleep is not a good option. Mammals evolved a chronic stress pathway to keep them going at times of prolonged activity, including flight or fight. However, ultimately, there is a natural end. Either the lions will tire and give up, or they will outrun our zebra and she will succumb to become their dinner. If she does survive this time, the zebra's chronic stress pathway can switch off again, normally within a maximum of an hour, and the body can return to a normal resting state.

Chronic stress is not inherently bad. However, in a domestic environment it is often not possible for the outcome to reach a natural conclusion within several minutes or maybe an hour or two. Hunter watches as his friends go out for a ride. He is a social animal; he needs the safety and security of his group. But he is left alone. He tries to follow them by barging at the door, but it stays bolted shut. He tries to call them back, but they do not return. He trots frantically around the stable, but he cannot get out. He is in acute stress. After several minutes, his body switches to the chronic stress pathway so he can maintain his attempts. He continues to shout loudly. He continues to rush at the door. He weaves frantically. His friends are gone for two hours. As they return, he is about to breathe a sigh of relief, but then his owner appears and puts on his head collar. She brings him out and sponges his sweaty coat down. He feels cold. She puts on his tack and climbs aboard. They go out for a two-hour hack of their own. He has to move quickly at times. He is alone, without his friends. When they finally get back to the yard, he is taken to a different stable. The farrier is coming. Hunter is fearful of the farrier because holding his legs up while the shoes are changed means he has to balance himself and this hurts his neck. He tried to communicate this to the farrier but he got shouted at to stand still. Shoeing takes nearly an hour. Hunter returns to his stable but now each of his friends are taken out of sight to the farrier, one at a time. This makes Hunter agitated that he cannot see them. Hunter is hungry; at this point he has not eaten for several hours and his stomach is still producing the constant acid that the equine's stomach has evolved to in order to receive trickle feeding and the accompanying natural acid buffer of saliva. His stomach hurts. The yard staff arrive with his haynet for the night; it is double netted with tiny holes to make sure it last for many hours. To get any hay out he has to grasp tightly with his front teeth and tug hard, which hurts his neck. He can only get little bits out at a time, he is hungry, this is really frustrating and painful for him.

In a domestic environment, horses often find that they are unable to escape from frustrations, unable to avoid potential threats, they cannot undertake actions that make them feel safe and they sometimes learn that no one listens to their important communications. They did not evolve for chronic stress to keep going for hour after hour, maybe day after day, sometimes week after week. This is when the chronic stress chemicals such as cortisol become a problem for the body.

In human medicine, there is a specific area of interest known as psychoneuroimmunology. It is a scientific study of how the brain and emotions interact with the nervous system and the immune system. Investigating the effects of long-term chronic stress is a large part of the research in this medical field. Behaviour and disease are even more intimately entwined than disease causing behavioural change, because behaviours, or at least, perhaps, the inability to perform certain species-specific behaviours, can actually lead, or certainly contribute, to disease.

Striving for balance

In earlier chapters, the role of maintenance behaviours was discussed. Every living organism strives for an internal balance. The complicated production and removal of chemicals, the internal temperature, the release of waste/breakdown products and the maintenance of internal fluid levels are absolutely essential to keep the organism alive. This necessity to find balance and the physiological processes that the body uses to maintain it, is called homeostasis. Each species of animal finds its balance in slightly different ways. The horse grazes for at least 60 per cent of a 24-hour period; it needs large quantities of low-calorie, high-fibre forage to maintain the balance in a complicated digestive system that has evolved so that this species can survive in places on the planet where plant growth is minimal. Horses are considered one of the most efficient grazing animals on Earth. Their ability to extract nutrients and calories from poor-quality plant material is amazing. So, over millions of years, they adapted this ability by developing special teeth for grinding, a stomach that is highly acidic to begin the breakdown of fibrous matter and a long and complicated intestinal system finely tuned to extracting maximal nutrition inaccessible to many other species. Balance for the equine comes from almost constant chewing and swallowing, from eating and moving almost continually, from extracting some nutrients from browsing; the eating of twigs, leaves and strips of bark. As a prey animal out in the open, they need to sleep but they can balance the brain chemicals and electricity that sleep seems essential for, in less than four hours in every 24, because to lie down and lose consciousness for long periods of time would not have resulted in successful predator avoidance. Sleep requires the herd to come close together and personal space shrinks considerably; some individuals can lie down and lose consciousness for several minutes but only because a few others are fully awake, acting as sentinels. This act is not altruistic as it is reciprocated; the sleeping individuals get up and take their turn to "look out" while the original observers get their rest.

Certain "core" or maintenance behaviours are the basis for achieving this internal balance or "homeostasis". According to equine scientist Waring, there are four main categories of maintenance activities: sleep and rest, ingestion, body care and comfort (Waring, 2003).

In 1988, scientist Marian Stamp-Dawkins coined the use of the phrase "inelastic demand", meaning that resources or environments that animals were prepared to work very hard for suggested their high value, and that the prevention or limitations on these "inelastic demands" was a concern for the animal's welfare (Dawkins, 1988). She describes how these behaviours are often the core activities affecting homeostasis, that they are crucial to survival (and reproduction ability) and how animals prioritise these essential activities above others. Prevention of these key activities is often referred to as "goal frustration" and leads to stress, which, depending on whether the animal can do something to change the situation and bring about the opportunity to undertake a similar activity to meet their need, or for how long the animal remains unable to obtain the "goal", can either be a fairly short-term experience or a much longer-term issue.

Although goal frustration is one of the most common causes of long-term chronic stress in the domestic horse, fear is also a frequent cause. Fear has been described in previous chapters, but with regard to veterinary and paraprofessionals, it is important to remember that many examinations, diagnostic procedures and both prophylactic (e.g. vaccinations) and treatment protocols, will often cause fear in the equine. Ways of minimising this will be discussed in the next section.

Longer-term chronic stress (that which lasts days, weeks or even months) ultimately affects the health of an animal. The evolutionary purpose of the glucocorticoids, as previously described, is to minimise energy-rich projects as part of a short-term adaptation to survive in the moment. When those stress hormones are present for prolonged periods, the projects remain switched off and a poorly functioning immune system, reduced tissue repair and decreased ability to sleep effectively all start to impact on the health and well-being of the individual. This is why it's important to be aware of the normal behavioural motivations of horses and also understand how they learn, so that we can avoid frustrations and fears that can result in long-term stress impacting on their health.

Getting back in balance is the body's priority. When an individual is subjected to longer-term stress, coping mechanisms are normally initiated to help try to "make the most of a bad situation". Animals who are undertaking coping mechanisms to deal with the stress, are often "surviving" but not "thriving".

What are coping strategies for the domestic horse? One of the first strategies many individuals apply is to reduce the number of "elastic" behaviours that they undertake. For the domestic horse, this might be activities such as play – both with objects and also with other horses; horses are a very playful species, offering a wide range of play behaviours throughout their lives, so the horse that isn't offering regular play with objects or other horses might be using this reduction as a coping strategy. Exploration of the environment is another elastic behaviour that might be reduced; horses are naturally curious and enjoy investigating objects, materials and other stimuli in their surroundings, so the equine who shies away from novel situations or investigating new things, might be suffering from chronic stress.

Some coping strategies revolve around increasing maintenance activities, such as eating and chewing more, drinking more and sleeping more; maintenance behaviours give the individual the opportunity to experience internal feel-good factors and these help offset the feeling of stress. Unfortunately, this strategy isn't as helpful as it might sound; homeostasis is not only affected by doing too little of an activity but also by doing too much. Imagine if you were to go from drinking a normal amount of water daily to greater amounts; your kidneys would have to work harder to get rid of the excess fluid they didn't really need. Similarly with food and with sleep. Eating or sleeping more than normal might not have any negative consequences for a few days, but after a week or two, you will start to gain weight and this puts a lot of extra pressure on many parts of the body, ultimately worsening stress, not improving it.

In some cases of chronic stress, animals start to develop abnormal behaviours as coping strategies. As described in Chapter 1, perhaps the most common abnormal behaviours we see in equines are stereotypies (e.g. weaving, crib-biting and box walking, although there are many more). Originally seen as bad behaviours, with a greater understanding of why horses develop and then sustain stereotypical behaviours we now appreciate that it is the environment of the horse causing, or historically causing, stress that triggers them, and that they are actually attempts at feeling better in a time when homeostasis is feeling compromised. As covered in Chapter 1, stereotypies are defined as "repetitive, invariant, behaviour patterns with no obvious goal or function" (Mason, 1991) and Mason describes how the performance of stereotypies in many individuals actually reduces the amount of glucocorticoids in their bodies, the main reason they are referred to as coping behaviours.

Despite the general acceptance that stereotypies are coping strategies which improve the animal's ability to deal with a stressful environment, there is also no doubt that stereotypies can cause physical problems for the individuals that carry them out. Remember Hunter who has been weaving frequently for many years, but such a repetitive, abnormal strain on the joints in the neck from this repeated action has led to the development of arthritis. Crib-biters are often much harder to keep weight on than non-crib-biters, mostly, it is believed, because they spend significant periods of time choosing to undertake the addictive behaviour rather than actually eating. Stereotypical box walkers may suffer from muscular strains or underlying joint issues from repeatedly turning sharply in the same direction. Therefore, although the development of abnormal behaviours such as stereotypies is undertaken by the animal as an attempt at coping, the detrimental side-effects are still a fairly high "cost", making such coping strategies "maladaptive" in the long term.

Photo: Suzanne Rogers.

This concern over the "maladaptive" elements of stereotypies means that the equestrian world is full of gadgets and housing designs to physically prevent the horse from carrying out the behaviour. There are anti-cribbing straps for crib-biters and wind suckers, anti-weaving bars for weavers and even air sprays and electric shock devices aimed at punishing the individual who carries out repetitive door banging or other compulsive acts. However, as Mason described, these undesired behaviours are coping strategies themselves, and physical prevention or punishing the horse for carrying them out is likely to add to the stress which is underlying the behavior, not help it. The horse may stop acting out the specific behaviour but is highly likely to develop an alternative coping strategy or, sadly, give up trying to cope with the stressors and allow the body to be overwhelmed with stress-induced side-effects, often resulting in significant ill health.

Having explored the links between behaviour, stress and disease, let us turn our attention to how vets and veterinary paraprofessionals can carry out their essential roles in a behaviourally-minded way.

Handling the horse – with behaviour in mind

Unfortunately, as far as the horse is concerned, vets are not exactly the "bringers of good things". At best, vets are likely to be neutral – they might turn up, stroke the horse, examine the horse, discuss the findings and leave. At worst, they can be a very negative experience – they might turn up, examine the animal causing fear and pain, give a painful injection or other treatment procedure or several, then leave.

Horses, like all mammals, learn from previous experience. They often form strong memories about past events, particularly those that were associated with fear or pain. They look for reliable predictors of things that cause these feelings, so that they can recognise them early, before the fear or pain can begin, and in a natural setting would then choose to avoid that situation. The appearance of the vet: a frequently smartly dressed individual, carrying a box of "tools", often wearing a stethoscope around their neck and smelling of medicines and cleaning agents, is a clear stimulus to pick out as a predictor of possible fear or pain. In the domestic setting of confinement in buildings or confinement by head collar or bridle, horses are often unable to escape these potentially threatening stimuli.

When an animal recognises a potential threat, their stress pathways already outlined, namely the sympathetic nervous system and the chronic stress pathway known as the HPA axis (hypothalamic–pituitary–adrenocorticol axis), are activated.

This process of switching on these pathways in preparation is referred to as "sensitisation". When an animal is sensitized, the result is the ability to move quicker and react faster; from an evolutionary perspective, the aim is to help the animal escape potential threat quicker. This means that just the arrival of a vet on a yard may initiate stress pathways readying the animal for fast action, even if that vet isn't actually there to examine them as an individual! Animals learn by pairing things together; in many cases, the arrival of the vet leads to an unpleasant outcome for the horse and they are unable to appreciate that the intention of the medication or other treatment is for their longer-term "good". When faced with such a stimulus predicting fear or pain, the horse would normally choose to walk or run away to avoid it, however, the prevention of flight by being in a stable, or being restrained on a head collar or bridle, means the horse needs to choose an alternative way of dealing with the stress. For many horses this is simply to "freeze": by standing as still as possible for several seconds or several minutes, they hope that the stimulus will simply go away – and for many individuals, this is what happens. However, it is important to recognise that when an animal goes very still during examination or treatment; their stress chemicals are still elevated. If there comes a moment when they feel they can no longer cope with just "freezing", the build-up of stress chemicals in their brain and blood at that moment is high and sudden attempts at flight or fight can be extremely fast, often knocking over the handler or vet, and with a significant risk of injury to all parties. Therefore, it is essential to recognise that "freeze" is not a sign that the horse is OK with the procedure being undertaken; failing to recognise freeze behaviours is probably the most common cause of physical injury risk when handling equines for veterinary examination or treatment.

Similarly, some individuals, on recognising that flight is not an option due to the confining situation they find themselves in, will turn to "fight" as their behavioural choice. Aggression is most commonly displayed when animals are in fear or pain, and when an animal is both fearful and painful, high levels of stress chemicals and a high desire for self-defensiveness as previously discussed, make the individual a high risk of injury to those around him.

How can we make the handling of equines for veterinary procedures safer for the handlers, the vet and the horse? There are four main methods that can be used: habituation, counter-conditioning, overshadowing or chemical restraint (ideally with the simultaneous use of anti-anxiety medications).

Habituation is the "waning of an original response to a novel stimulus as the result of gradual and repeated presentation". The difficulty with habituation is this term "novel", meaning that if the stimulus is not a novel one but one the animal has met before, previous associations may have been made. When we learn about a stimulus or have an experience, we form long-term memories. If we meet a stimulus and that initial experience is a negative one, this type of long-term memory is stored in such a way that it can never be forgotten. At best, if not replayed or reactivated in the brain by a repeated experience or a similar experience, then the memory can become weak: think of a language or musical instrument you learnt in your early school years that you have not used for a long time. You may think you have totally forgotten how to play that instrument or how to construct a sentence in that language, but actually you still have the memories of how to do it, they are just weak. Once you start to practise again, you will gain the ability to play or speak much faster than it took when you originally learnt. Memories, we say, are "gone, but not forgotten".

True habituation requires the exposure of the horse to a novel stimulus or situation; in a gentle and gradual way so that without experiencing fear, the individual can learn that the stimulus or experience is neutral and does not need to be feared. In dog training, a lot of attention is paid to habituation – exposing puppies to lots of different stimuli, gradually over several weeks and then ideally across their lifetime so that the concept of meeting new things and having new experiences remains a positive one. Sadly, in the equine world, habituation is not given nearly enough attention.

In some circumstances, horses are actually experiencing the opposite effect of habituation by initially having that novel experience paired with restraint, fear or even punishment, resulting in sensitisation or even a classically conditioned, automatic, reflex reaction of fear without conscious input. Examples of this could include "imprint training" where a very young foal is physically restrained (a significant cause of fear in such a young, vulnerable prey animal) and is exposed to several stimuli such as having their nose, mouth and ears handled, insertion of thermometers into the rectum, clippers, being rubbed all over with plastic, even the application of a small saddle. Being unable to escape a fear-inducing stimulus is referred to as flooding in behavioural terms and does not result in the stimulus being perceived as neutral and non-fearful.

It is possible to habituate the equine to stimuli such as the smell of the vet, being touched on the neck, having the mouth gently handled, having bandages applied, and so on, but it needs to be undertaken in a careful, gentle, non-fear-inducing way, normally in the young animal, for true habituation to occur. In the majority of domestic equines, habituation has not been carried out successfully so we are left with an animal that already has fear associations with a range of handling and veterinary stimuli, leaving us with other options should we wish to minimise stress during the veterinary procedure.

As introduced in Chapter 3, counter-conditioning normally refers to a "retraining" programme used when an animal has developed a conditioned fear response to a stimulus so that the sight or sound or smell of the stimulus immediately elicits switching on of the stress pathways and a fear behavioural response. This type of learning is often referred to as Pavlovian classical conditioning, given that it was Ivan Pavlov who first discovered and recorded it back in the 1890s – many of us have heard of the original experiments where Pavlov paired the ringing of a bell (a previously neutral stimulus) with the sight and smell of food to hungry dogs so that the digestive reflexes were unconsciously switched on in the body including salivation. Repeated pairings led to the dogs showing salivation just in the presence of the ringing bell – even when no food was present at all. The dogs had learnt that the bell was a reliable predictor of the food arriving and the unconscious initiation of the digestive reflexes began upon hearing it. The salivation in the presence of the bell became known as a conditioned response. In the same way, the arrival of the vet on the yard often initiates the flight or fight fear reflexes in the body as their arrival is a reliable predictor of having a fearful experience – normally a potentially painful examination or a needle. Horses often show conditioned responses – reflex, unconscious reactions – on the vet's arrival, such as starting to jog on the spot, fidgeting about or even quite marked attempts at flight, even before the vet has physically approached them.

Counter-conditioning refers to a new learning process whereby, in this case, we attempt to replace the switching on of reflex fear pathways with the reflex production of calm, "rewarding" pathways. One of the important things to remember when attempting to counter-condition is that the presentation of the conditioned stimulus immediately triggers the unwanted internal chemicals. This means when we begin a counter-conditioning programme we have to first be clear on exactly which stimuli are to be presented. Then we must be clear on how they are presented, for example, how close/ far away from the horse, whether and how the stimuli move, whether there is noise and if so how much, and in which environments they are shown. We then break down the stimulus into more "unrecognisable" pieces; for example, we might start with the smell of a mild antiseptic on a towel out on the yard, rather than a strong antiseptic smell being on clothes worn by a human approaching the stable, and reward the horse for demonstrating calm behaviours. The reward element is key as we are intending to replace the reflex feelings of fear with reflex feelings of pleasure – this is why the training is referred to as "counter", literally meaning "opposite" conditioning. Once we have introduced the horse to the stimulus broken down into smaller pieces, we start pairing the pieces back together again – for example, the smell of antiseptic on a towel, present at the same time as a person resting their hand on the horse's neck muscle (a common site for intramuscular injections) and rewarding the horse for remaining calm.

Counter-conditioning can be extremely effective. However, the brain now has two sets of memories about the stimulus: the original memory that the stimulus, for example the vet, approaching, was scary, and now a second memory that the vet approaching can be pleasurable. It can be difficult to maintain the easiest memory for the horse to choose as the second one, when vets frequently arrive and undertake a scary, pain-inducing experience. It is essential to remember that counter-conditioning will be rapidly unlearnt if the vet becomes the predictor of bad things again – just like when Pavlov stopped offering food to the dogs in the presence of the ringing bell; eventually they stopped salivating as the bell no longer reliably predicted food requiring digestion. This process is called extinction.

One of the ways to avoid extinction once we have counter-conditioned a horse to the vet, is to use "overshadowing" for future vet experiences, as well as practising some of the counter-conditioning programme in between vet visits. Overshadowing is the term used to describe presenting more than one stimulus at the same time to an animal, but that one stimulus is more valuable to that individual so distracting them from paying attention to the other stimulus/stimuli. Think of the following example for yourself: you are having dinner in your favourite restaurant and have ordered a new dish; it has just arrived and you are finding it to be the best-tasting plate of food you have ever eaten. Your friend who you are dining with starts telling you a pretty boring story about something that happened to them yesterday; which do you think you'll remember later on that day? How amazing the food tasted including how it looked and smelt on the plate, or will you remember a bit of uninteresting information that had little relevance to you?

Overshadowing doesn't have to involve stimuli that are polar opposites in terms of being amazing or being unpleasant. It just means one stimulus is more "valuable" to that individual. For example, when horses are restrained for veterinary procedures such as having a blood sample, it is fairly common for people to hold a handful of the horse's neck skin – often referred to as a "skin twitch". If the discomfort or fear induced by having someone hold tightly onto a handful of your skin is more "valuable" to the horse than the needle going into their vein, then this is also overshadowing – they are both unpleasant experiences but the skin twitch is more "valuable" than the needle and the horse may well not have even noticed the needle going in. In training terminology we normally refer to this "value" as "salience" – we would say the skin twitch is more salient than the blood sample for this individual. However, if we use fear-inducing or painful stimuli to overshadow other fear-inducing or pain-causing stimuli, such as someone might twist the horse's ear by its base – colloquially known as an "ear twitch" – to overshadow a needle being put in the horse's leg as part of an investigation into the cause of lameness, then the vet still remains the reliable predictor of scary, potentially painful things. Overshadowing using salient, negative stimuli such as skin, upper lip and ear twitches is a fairly common technique used in equine diagnostics and treatment, and it can be pretty effective at getting procedures done, however, it only goes to teach the horse that vets really are people to be avoided, or even attacked, which in the long term makes the relationship between vets and their patients increasingly difficult.

Overshadowing can be achieved using things the horse finds particularly pleasant rather than relying on things the horse particularly dislikes. Examples can be, knowing that a horse has a particularly itchy spot under their chest that they really enjoy having scratched, and then the handler scratching that itchy spot hard at the same time as the vet gently inserts a needle for the blood test. Most horses are pretty keen on some sort of tasty food – the key when using food to overshadow is to ensure the food remains their entire focus, so the handler should hide the food in their hand and have the horse search hard for little tasty pieces while the vet carries out the procedure, rather than offering some food in a bucket, which often results in the horse grabbing a mouthful and then looking round to see what the vet is doing!

Used well, overshadowing can be a highly effective tool to prevent stress pathways being triggered, to help keep the horse physically still and calm for certain procedures – particularly those involving needles – and for preventing the horse from learning negative associations about the vet, including once a counter-conditioning programme has been completed but now future events such as blood samples or other injections need to be undertaken.

Overshadowing has its main limitations when: we need to do something in the mouth of a horse such as dental work where giving the horse food simultaneously is not an option; the horse's fear of the vet or procedure is too salient so that any other stimuli including tasty food are too minor in comparison; or when the

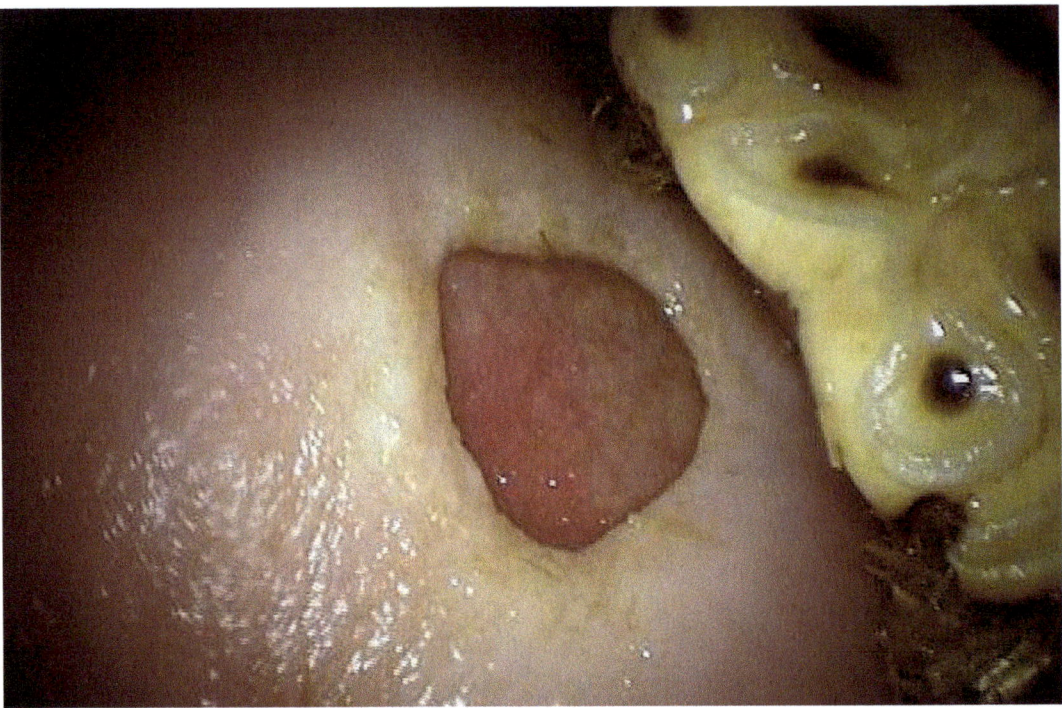

Figure 8.2: Mouth ulcer caused by dental disease. Photo: property of The Equine Dental Clinic, UK. Image provided by Mr C Pearce MRCVS of The Equine Dental Clinic, Dorset, UK.

procedure is going to take a long time, such as an operation or an imaging process like computerized tomography (CT) scanning or gamma scintigraphy.

In those situations, the use of the fourth option given above, chemical restraint, might be the best option. Chemical restraint or sedation is a good choice for the diagnostic and sometimes treatment elements of the veterinary intervention as it minimises the physical risk of the horse to the handler and vet. It is essential to remember, however, that even well-sedated horses can have sudden surges of stress chemicals, particularly adrenaline, resulting in sudden flight or fight attempts.

Chemical restraint can be administered to the horse using three methods: oral (usually in the form of a paste such as acepromazine or medetomidine); intramuscular (such as medetomidine or medetomidine/opioid combination); or intravenous (such as xylazine or medetomidine or romifidine, often with an opioid combination). Each method has its pros and cons as outlined in Table 8.2.

Whichever method is used, it is important to remember that sedation does not result in loss of consciousness. The phrase "chemical restraint" gives the more accurate description as an animal is less able to move as the result of the drug administration. Animals that have been sedated are still conscious and therefore aware of stimuli and events in their surroundings and those that they are directly experiencing. This is particularly important to remember when painful procedures are being undertaken such as minor surgeries including wound stitch-ups, as the horse will be able to feel the pain but not necessarily be able to physically do anything about it. The same applies for procedures such as mane pulling or clipping where the horse may be experiencing pain particularly in the case of clipping, if techniques are poor

Table 8.2 Comparison of administration methods for chemical restraint with behaviour in mind

Administration method	Pros	Cons
Oral	Can be undertaken by the owner/handler before the vet's arrival therefore reducing stress before the vet is recognised by the horse Non-painful method therefore beneficial for horses with marked injection or touch reactions Often cheaper than other methods Can be given in food fed from a bowl in products designed to be absorbed from the stomach	Not well-suited to horses that get agitated by having their mouths handled (for example, if they have an aversion to pastes due to prior experience with wormers) Certain products must be given under the tongue and not swallowed to be effective and this can be difficult to achieve practically, particularly in horses that get easily stressed by handling around the head and mouth Difficult to overshadow with food Take up to one hour to work making timing of sedation difficult to predict and hard to "top up" should the dose not be quite as intended
Intramuscular	Useful for horses with learnt aversions to intravenous injections as there are multiple sites across the body suitable for muscle injections Works moderately quickly, drug-dependent but normally within 15 minutes of administration Fairly easy to overshadow needle placement as the head is free to receive food Can be given "from a distance" or from behind a barrier for horses that are aggressive or physically dangerous with flight responses, as the site of injection is large and the procedure not as anatomically "accurate" as entering a vein	Not so easy to tailor to the required depth of sedation as the individual animal's absorption rate from muscle tissue is less predictable. This can result in poor sedation level or even far too heavy sedation The restraint of the horse and the pre-injection procedure of numbing the skin by banging with the hand or using a local skin twitch, creates an obvious predictability for the horse to which many horses become classically conditioned and begin reacting before the needle is placed Cost – higher amounts of drug are required due to slower absorption from muscle tissue compared to intravenous administration, normally making this the most expensive form of chemical restraint
Intravenous	Works within minutes of administration Easy to tailor to the horse's size and to "top up" should depth of sedation not be as required due to speed of absorption Fairly easy to overshadow needle placement as the head is free to receive food	The preparation of the jugular vein for administration, e.g. clipping hair if required, wiping clean if required, the raising the vein before placing the needle, creates a predictability for the horse that often leads to a classically-conditioned fear response before the needle is placed Typically, the most accessible vein for needle placement is the jugular meaning that only two locations on the body can be used (left and right jugular), again providing an obvious predictability for future learning

and rushed, and yet this is often overlooked by the handlers.

Sedation drugs can be combined with pain relievers, some of which, such as opioids (morphine-like medications), might have some mild sedating properties of their own (interestingly, in the equine, opioids given on their own tend to induce a more excitable state, yet when combined with a second sedating medication, help to reduce the total dose required and increase the sedation depth as well as providing a general reduction in pain perception). Again, this is beneficial to the animal undergoing a painful or uncomfortable procedure, and may reduce the overall perception of pain intensity, but does not necessarily remove all the pain experience. The use of local anaesthetics to totally numb tissues and cause 100 per cent blocking of pain recognition should be considered in as many procedures as possible.

As the sedated horse is still conscious and can experience fear and pain but is often less physically equipped to do anything about it, chemical restraint should be considered as a possible act of "flooding", as previously described. It is essential for the welfare of the horse and for future learning that gentle and respectful handling occurs during chemical restraint and is combined as above, with the use of the most effective pain reduction possible. Some sedation drugs have an anxiety-reducing effect in the body and these sedatives should be used in preference to those without that known property where possible. In other species, including humans and dogs, the class of medications known as benzodiazepines is the most commonly accepted class of sedatives having anti-anxiety effects and in people is also recognised as having "memory blocking" effects so that administration of this drug prior to a potentially stressful procedure can actually reduce the learning that may occur at the time. However, benzodiazepines are not routinely used in equine medicine as their administration often causes a quite marked "ataxia" – the animal being unable to coordinate their balance effectively – making these drugs potentially dangerous to the animal itself and those around him. Benzodiazepines can be used in smaller equines where administration of chemical restraint can be accompanied by the safe lying down of the animal and physical assistance in re-standing is possible, for example in foals and Shetlands, but their suitability for larger equines, where remaining on their legs is necessary, often renders this category unsuitable for use.

Alpha 2 agonists such as detomidine or romifidine, are believed to have some anti-anxiety properties so these are the most suitable drugs for sedation accompanying veterinary procedures or even sedation for other fearful experiences such as difficult loaders, for those fearful of the farrier or horses with fireworks phobias.

Treatment of medical problems and behaviour

Administration of medications

We have already considered some of the behavioural aspects of medication administration in the previous sections, including the use of overshadowing and the pros and cons of chemical restraint techniques. Preparing the horse for possible future events is always a benefit and therefore habituation or counter-conditioning programmes for handling the horse's mouth in order to administer oral solutions, is a good idea. Once these programmes have been followed, the intermittent practice of allowing the horse to lick nice-tasting pastes off a washed-out worming tube or a large plastic syringe, is a great idea to keep the horse in a positive mindset about having solutions put into their mouths. Consider using honey, or yeast extract or perhaps home-made, sticky, palatable mixtures using small amounts of icing sugar mixed with water and a flavouring that is favoured by the individual equine.

Similarly, it can be worth practising adding flavoured powders to small portions of damp chaff or other moist feed substances (e.g. a handful of soaked beet) to keep the horse accepting of novel flavoured powders being intermittently given. You can try adding a tablespoon of plain flour to a "treat" feed for, example, not their regular meals, and adding additional flavours such as dry herbs or spices as variety to help the horse accept that novel flavoured powders can be a good thing. This helps create a positive mindset should something like antibiotic or anti-inflammatory powders need to be given as part of a treatment protocol. Ensuring the horse gets variety in his/her diet in the form of different flavoured hays (try soaking a small net in water infused with herbs or spices such as chamomile, tarragon, cinnamon, bay leaf – options are wide ranging!) or different live or dried plants chopped up and mixed through forage on the ground (you can try dandelions, dried nettles, parsley, coriander, willow, juniper berries, rosehips), can help prevent neophobia – the fear of trying new things. Horses that are used to experiencing different flavours and textures are far less likely to reject medicines and it will also be easier to "mask" medications using their known favourite strong-tasting flavours such as turmeric, liquorice, spearmint, and so on.

Box rest and reintroducing movement, with behaviour in mind

Many injuries (such as wounds, tendon strains or fractures) and diseases (such as laminitis) will require a period of immobilisation to ensure maximal healing and then this is normally followed by controlled and progressive mobilisation. Strict rest is often the prescribed immobilisation for acute injuries in the equine. However, total confinement brings with it a high degree of goal frustration as previously discussed, as well as fear from being unable to flee potential threats, resulting in stress. There is a challenging balance in meeting the physical needs associated with best healing opportunities versus preventing psychological issues that might result in physical reactions, such as excess box walking, which oppose the benefit of the confinement. There is also a need to consider how frustration and fears result in stress, of which suppression of the immune system and reduced tissue healing capacity are undesirable side-effects.

Is it possible to meet the psychological needs of the equine while physically confining them? The answer is that whilst any physical restriction for an animal who has evolved to use flight as their main form of self-defence will always create some element of stress, this is affected by the animal's individual previous experiences and learning, their temperament and the way the situation is practically managed.

Key aspects to consider:

- *Companionship*: Horses are highly social beings and rely on living in a group to feel safest as a prey species. In many cases, the confined area could be used to house the injured individual *and* his friend so they have direct contact with one another at all times. If it is deemed necessary to remove the companion for periods of time, for example for some grazing time or exercise, then the author has always found that leading both individuals out of the confined area together, then "overshadowing" with something such as returning into the confinement area for scattered feed as the companion is led away, is far less likely to induce a marked stress response compared to leading the companion out of the confinement and leaving the "patient" stuck behind a barrier. Ideally, an alternative companion would be provided for the period the main "friend" is away, even if this is just tied up outside the confinement area with a haynet or similar.

Where sharing a direct space isn't suitable, ensuring the patient is confined adjacent to his main companion and with some physical access for reassurance, grooming and over-barrier play, is the next best option. This can be achieved using stables that have just a mid-neck-height wall between them, or stables with bars separating them. Daily periods with the main friend tied up outside the stable door, ideally with a chain rather than a door present to allow freer access for physical contact, is preferable where suitable. Again, it is important to consider how any separations are handled: the use of overshadowing with enrichment opportunities is the most ideal (see section below) or where appropriate, moving the patient out of the stable for hand grazing or "station-led" hand walking (see below) can also be invaluable in preventing highly active stress responses from the confined individual.

- *Forage and browsing*: All the feed enrichment ideas discussed in Chapter 1 are relevant. Ad libitum access to low-calorie, high-fibre forage, normally hay, is ideal for the equine on box rest, with the exception of those individuals with calorie-restrictive diets such as laminitics. It needs to be remembered that horses have evolved to eat 80–90 per cent of their diet directly from the ground and that this benefits them in several ways, including maintaining a healthy respiratory system. Placing forage along the walls, rather than just in one corner, or at least using all four corners, encourages gentle moving and eating, appropriate for almost all conditions except those in which total immobilisation is a necessity – see managing the cross tied horse, below. Straw beds can also be useful as horses will top up their "grazing" requirements from this lowest-calorie bedding, however it must be considered that the lack of movement in the 24/7 confined horse does predispose the increased risk of impactive colic.
- *Sleep*: Horses, like all mammals, need adequate sleep to ensure the healthiest immune system and most effective tissue healing. However, whilst horses can get some slow wave sleep while standing up, in order to get the best quality slow wave and rapid eye movement sleep, horses do need to lie down. Horses prefer to lie down and will get better quality sleep in the presence of a trusted companion that can act as sentinel, and if they can see their surrounding environment whilst lying down (compared to seeing nothing if confined within solid walls). Is it possible to confine the horse in a field shelter set up with a gate, or bars across the open front, so they can see companions easier and also see outside whilst lying? Can the stable have a small post and rail pen attached so the horse can move freely between the pen and the stable? Depending on the time of year, the pen may be soil so lying down is comfortable outside, or the pen can be on concrete but have rubber matting or even something like bark chippings or sand as a surface, making lying in the pen an option. As above, at the very least, can the patient have periods of the day where the favoured companion is outside their door, in effect acting as sentinel?
- *Making choices and exploring*: Prolonged confinement is not only frustrating but also

very dull. The natural horse's environment is full of stimulation and choices. This is possible for the horse on box rest; multiple options have been given above, but also consider using a cardboard box with grated or peeled veg hidden among soft toys, large rubber dog toys or screwed-up balls of paper so the horse has a "toy box" to play with that will not cause injury should they put a hoof in it. There are a number of enrichment toys available on the market but it is important to consider how frustrating they can be if made too difficult – in a natural environment, horses do not need to "work" to acquire food and different tastes. Investigating new objects, such as different-coloured hand towels or different-textured cushions, or sniffing scented cloths made with handkerchiefs sprinkled with one or two drops of essential oils (this is not intended as part of a complementary treatment for their condition but as just an enrichment opportunity for sniffing and investigating), can all be safely done in very small set-ups if undertaken sensibly. Rachaël Draaisma is one of the first people to recognise the olfactory potential of the horse and its use in enrichment, some of which can be applied to the confined patient (Draaisma, 2017).

- *The cross tied equine*: Occasionally, a horse has an injury or disease that is so severe they need to be restricted from any movement, including lying down, for several weeks. During this time, many of the above principles can still be applied, however, consider that instead of scatter feeding on the ground, this can be done off straw bales stacked to chest height outside the stable door/chain, and some head down time daily can be undertaken if a handler is present for several minutes to stand and hold the horse *in situ* on a lead rein with ground forage present. "Table top" games can be played, allowing the individual to explore flavours, objects and smells on a chest-height bale or barrel, or even teaching simple "selection" games using reward training, such as, if the handler shows a picture of a circle, can the horse select the circle from one of three shapes on the table by touching it with their nose?
- Sleep deprivation can be an issue for the cross tied horse so consider putting a mattress or padded bales against the wall on one or both sides, so that there is a solid, slightly soft, surface to lean against to get more effective slow wave sleep and several seconds of REM sleep before the physiological loss of stabilising apparatus kicks in.

Introducing movement after confinement

Proprioceptive rehabilitation is believed to enhance the nerve and muscular control mechanisms so that joints and associated soft tissues have better stabilising properties when the limb or body is actively moving. Overall, it is accepted that specific exercise programmes, including slower, controlled activities that allow for a wider range of nerve and muscle control, are an essential part of a multimodal approach to rehabilitation, alongside appropriate nutrition, the use of anti-inflammatory and "tissue-enhancing" medications (such as cartilage-supporting agents) and appropriate pain relief (Goff, 2007).

Unfortunately, there is often a marked "rebound effect" following a period of confinement that makes the patient want to "explode" as they leave their stable, which can make the transition from total confinement to controlled in hand exercise impractical, if not even dangerous for handler and horse. Although the use of sedatives is a possibility, most sedatives at doses that are high enough to limit the "rebound" activity desire, also cause reduction in gut activity, potentially increasing colic risks, and can cause balance issues making the animal less stable on its limbs and a further risk to itself or handlers. The author has used two methods with a good level of success to reintroduce hand exercise to the horse:

Station-focussed walking

This method requires setting up five or six "stations" around the area the horse is to be hand walked in – ideally the yard adjacent to the stable, initially. These stations can simply be weighted buckets (such as rubber feed buckets or plastic trugs with a small log or similar in them) or can vary between buckets and "mats" such as smooth, material, household doormats (rubber-backed ones are often easier to avoid being moved by the horse or the wind during the process) or towels. They need to be placed in an obvious location so they are easily noticed by the horse and a minimum of several metres apart but no further than 20 metres, at least initially. Each bucket should contain a small handful of something edible such as damp chaff, grated vegetables or high-fibre pellets. The handler should wear a jacket with a pocket that is full of replacement food. The first bucket should be placed just a few metres from the door, so that the aim is to teach the horse to walk out of the confined space slowly and within metres, stop and head lower to eat from the bucket. The horse is allowed to eat all of the contents, which should take up to 30 seconds, and then should be led onwards to the next station. As the handler and horse are about to leave station one, the handler should take a few pieces of food from their pocket and drop it into the first station without the horse noticing, this way, when they return to that station, there is food immediately available again. The horse and handler walk between stations, if required they can change direction of the circuit once they reach station one again. The aim of this process is to teach the horse that hand walking is an active but calm process with frequent, calm pauses. Normally, within several sessions, the number of stations can be reduced or the circuit size increased with bigger gaps between stations to allow for longer "active" walking periods between stations. Over a couple of weeks, stations can be phased down to three, in a more triangular shape and the horse can be taught that we skip out each alternative station, thus maximising the activity of the walking.

Target stick walking

In horse, dog and even cat reward-based training, the use of a target stick is relatively common. The animal is taught to touch the end of a stick to earn a small food reward. This training can even be a simple form of stimulation while the horse is on confined rest, however, it should be noted, as already mentioned, that horses did not evolve to have to work for food, so this can be frustrating for them; combined with the frustration of stabling, many stabled horses can find reward training quite stressful. It is therefore important that the horse is not hungry (for example, due the next batch of forage) at the time of training and also that a good shaping programme is in place, perhaps alongside a reward-based behaviourist or trainer's created programme for that individual, to ensure the training goes smoothly and without stress for the horse.

Taught well, target training can be a useful tool to use when introducing hand walking after a period of confinement. The target is carried by the handler and, in a similar way to station training, the handler presents the end of the stick towards ground level for the horse to reach and touch, on a regular basis to keep the horse focussed and calm during the walking activity (Figure 8.3).

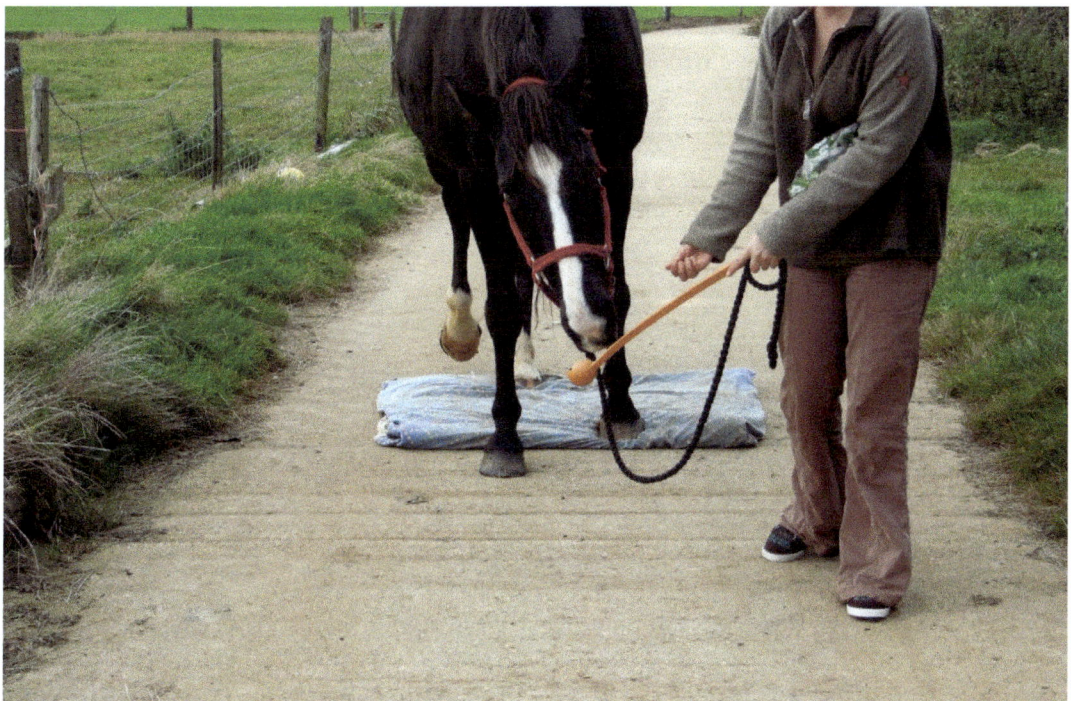

Figure 8.3: Target stick training can be useful in teaching the horse to hand walk in a calm manner. Photo: Amber Batson.

Target training is also useful to help the horse carry out physiotherapist-recommended stretching exercises, or to reward the horse for walking over different surfaces of varying substrates, cambers or even over objects with mild movements, such as mini-seesaws created with sheets of thick, non-slip ply balanced on a thin tree stake as part of proprioceptive retraining or core muscle exercises useful in certain injury rehabilitation.

Both station training and target stick walking can be valuable tools to teach horses to walk calmly back to their confined spaces following periods of hand grazing too.

A multi-modal approach to the horse with behaviour and medical problems

Let's return to Hunter. What did we do to help him with his recent development of aggression, long-standing stereotypy of weaving and his neck pain and associated allodynia and hyperaesthesia?

The first thing was to address the pain associated with the neck arthritis. Hunter was started on a long-term, anti-inflammatory and pain-relieving medication given as granules in his feed. He also began regular physiotherapy, carried out under "chemical restraint", which combined sedation, pain relief and anti-anxiety properties to maximise the effectiveness of his treatment. The sedation was given intravenously by the vet whilst the owner had Hunter search for flakes of horse feed in her hand (overshadowing) and was done in the presence of a calm companion. The physiotherapy included laser therapy as an additional pain-relieving technique and stretches to relieve his secondary muscle tensions. Hunter's owner was able to organise moving Hunter and his favoured companion into a quiet side yard with only two stables, and Hunter's door was left permanently open so that Hunter could move in or out as he chose; a hay box was placed near the door of his companion, as well as ad lib hay in the corners of his stable, so that Hunter no longer had to tug on a haynet and could eat in social contact. When Hunter's companion was taken out without him, Hunter was moved onto the main yard and tied up with haylage in a hay rack fixed at chest height so he did not get distressed as his companion was removed, and so he had social contact whilst they were away. These changes markedly reduced Hunter's weaving behaviours, which further reduced his neck pain. A combination of this stress reduction and pain relief resulted in Hunter no longer being aggressive and no longer having issues with his rug being put on.

Final thoughts

There are several ways in which disease can affect behaviour and behaviour can affect disease. Veterinary examinations and procedures for both diagnostic and treatment purposes as well as prophylactic interventions are likely to affect both the short-term and the long-term behaviour of the horse. Where injury or disease requires confinement as part of the treatment plan, frustration of goals and anxieties caused by reduction in flight opportunities and social interactions can both cause stress that may affect short- and long-term behaviour as well as limiting the effectiveness of the confinement itself. By understanding equine behaviour principles and how to practically apply these, we can improve the outcomes of veterinary interactions, improving handler and equine safety, reducing stress and limiting long-term consequences such as phobia development. Behavioural problems are a common cause of relinquishment of ownership in the equine world, alongside a common reason for the animal's euthanasia; delving deeper into the possible pain and disease components of these problems is a key element in improving equine welfare.

References

Breuhaus, B.A. (2011) Disorders of the equine thyroid gland. *Veterinary Clinics of North America: Equine Practice, 27*(1): 115–128.

Dalla Costa, E., Minero, M., Lebelt, D., Stucke, D., Canali, E. and Leach, M.C. (2014) Development of the Horse Grimace Scale (HGS) as a pain assessment tool in horses undergoing routine castration. *PLoS one, 9*(3): e92281.

Dawkins, M.S. (1988) Behavioural deprivation: a central problem in animal welfare. *Applied Animal Behaviour Science, 20*(3–4): 209–225.

Draaisma, R. (2017) *Language Signs & Calming Signals of Horses: Recognition and application.* CRC Press, USA.

Dyson, S., Berger, J., Ellis, A.D. and Mullard, J. (2018) Development of an ethogram for a pain scoring system in ridden horses and its application to determine the presence of musculoskeletal pain. *Journal of Veterinary Behavior: Clinical Applications and Research, 23*: 47–57.

Goff, L. (2007) Equine Treatment and Rehabilitation. In McGowan, C., Goff, L. and Stubbs, N. (Eds), *Animal Physiotherapy: Assessment, treatment and rehabilitation of animals.* Blackwell Publishing, New Jersey, USA.

Horseman, S.V., Buller, H., Mullan, S. and Whay, H.R. (2016) Current welfare problems facing horses in Great Britain as identified by equine stakeholders. PLoS ONE 11(8): e0160269. https://doi.org/10.1371/journal.pone.0160269

Horseman, S.V., Whay, R. and Mullan, S. (2016) *Horses in our Hands.* http://www.worldhorsewelfare.org/survey-equine-welfare-england-and-wales (accessed February 2018).

Mason, G. (1991) Stereotypies: a critical review. *Animal Behaviour, 41*(6): 1015–1037.

McCue, P.M., Roser, J.F., Munro, C.J., Liu, I.K. and Lasley, B.L. (2006) Granulosa cell tumors of the equine ovary. *Veterinary Clinics: Equine Practice, 22*(3): 799–817.

McDonnell, S.M. (2008) Practical review of self-mutilation in horses. *Animal Reproduction Science, 107*: 219–228.

Mullard, J., Berger, J.M., Ellis, A.D. and Dyson, S.

(2017) Development of an ethogram to describe facial expressions in ridden horses (FEReq). *Journal of Veterinary Behavior: Clinical Applications and Research, 18*: 7–12.

Pirie, R.S., Mayhew, I.G., Clarke, C.J. and Tremaine, W.H. (1998) Ultrasonographic confirmation of a space–occupying lesion in the brain of a horse: choroid plexus papilloma. *Equine Veterinary Journal, 30*(5): 445–448.

Waring, G. (2003) *Horse Behavior: The behavioral traits and adaptations of domestic and wild horses, including ponies* (Noyes Series in Animal Behavior, Ecology, Conservation, and Management). William Andrew, New York, USA.

Chapter 9

Working animals

Suzanne Rogers

Approximately half of the world's human population is still dependent on the power provided by draught animals, a staggering 100 million of which are equines. With entire extended families often dependent on the working capacity of just one animal, human welfare and animal welfare are inextricably linked. Only decades ago, it was thought that most of our coffee was cultivated with the use of working animals, at some point carried on the back of a working animal or drawn in a cart behind a working animal. This is unlikely still to be the case but illustrates the complexity of global primary resource production and trade, and how little we really know about the contribution of working animals, including equines. This chapter explores the intricate relationship between some of the world's poorest people and their equines, and why the welfare of these animals is crucially important, not only for the health and survival of those animals but also for the livelihoods of the people who depend on them. Given the abject poverty of the owners, is meeting the behavioural needs of working equines an inaccessible luxury? Is it possible to "work" equines with behaviour in mind? What are the components of a behaviourally-minded project aimed at improving the welfare of working animals? Instead of a story with the equine at the centre, for this chapter we will consider an owner named Samnang.

An owner called Samnang

Imagine you are a Cambodian pony owner called Samnang. Ever since you can remember, your day has followed pretty much the same sequence, at first watching your parents but then taking on more responsibility as you get older: get up early, take water to the little chestnut-coloured pony that is in a fenced-off enclosure underneath your raised house, feed him, harness him up, take him to the waiting place – a corner of a busy street perhaps where a long line of owners are waiting for work and will do so for hours. It's boring; you're holding a stick and you absent-mindedly swing it around, hitting the pony, because that's what you've always done. Eventually, you are approached by someone who has a job for you: a short distance away is a building site; you have to take rubble from a store to the site and come back again rapidly because if you do it quickly, you'll get back to the work queue faster, increasing the chance you will get another job that day. You load the cart; lifting each block of stone causes an old back injury to flare up so you throw them in the cart rather than place them, not noticing your pony flinch as each one lands. You are in a hurry, so have no time to balance the load or check the tyres are the right pressure. You drive the pony fast, at canter on the road, to meet the client's demands. You can hear from the uneven footfall on the road that your pony is starting to become lame, but you can't think about that now – rest later when there is no work, you need this job to feed your family as food has been scarce the last few days. A few more jobs later and after loading and unloading many times in the sweltering heat, you have time to rest. Aching, you seek out some shade, leaving your pony in the sun before trying to find some rice and meat for your family and perhaps later taking him to the river to bathe and have a drink. Then the next day you do the same again …

You have never really been to school – for example, the 2016 UNDP Cambodia Human Development Report (UNDP, 2016) states that the average Cambodian child receives 4.7 years of education and around one in 25 children will not live to the age of 5. You have always worked like this. You *know* that your pony needs more food, more frequent shoeing, that the harness made from an old motor bike tyre needs to be repaired and that the lesion on his shoulder is unlikely to heal if you keep using him for work, but those things will have to wait; the need for work is pressing and there isn't much you can do.

This is an example of the reality for millions of people around the world and the horses, donkeys and mules in their lives. We will revisit Samnang at the end of the chapter.

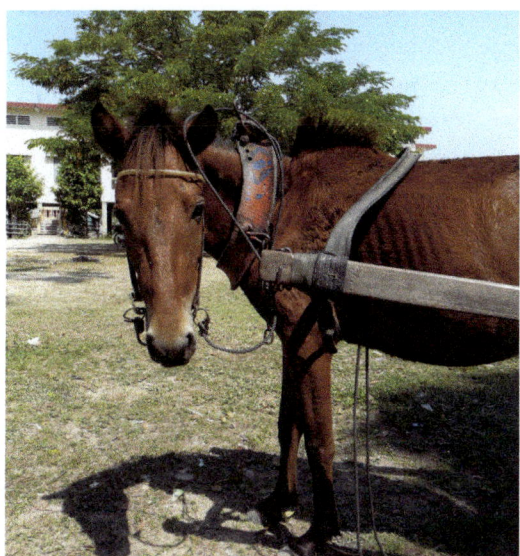

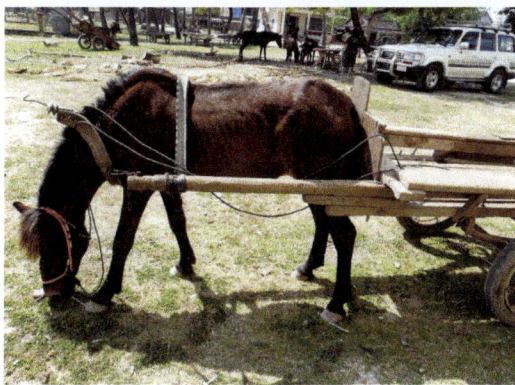
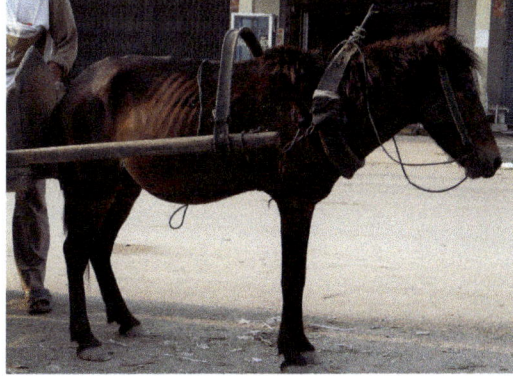

Figure 9.1a-b-c-d: a and b, typical Cambodian working ponies; c, ponies given opportunity for social eating; d, the lack of breeching can clearly be seen; this example does not have padding on the front of the cart. Photos: Suzanne Rogers.

Working animals 185

Numbers and distribution of working equines

It is estimated that there are approximately 100 million working equines in low GDP countries helping around 600 million people globally, very often in poor and marginalised communities (reviewed in Brooke, 2015). Population figures for animals are notoriously difficult to calculate but even allowing for a significant error, these animals clearly still represent a vast power resource. (For a comprehensive consideration of the numbers of equines worldwide, see the review by Clarkson, 2017.)

It is estimated that 80 per cent of the world's equine population are found in low GDP countries and that half of the world's donkey population, almost all of which is used for work, is found in Asia, just over one quarter in Africa and the rest mostly in Latin America, mainly Mexico. You might think that the use of equines for pulling carts or carrying loads is decreasing as motorised vehicles become more common. However, although a decrease in equine populations has been seen in some countries, the number in Africa, South Asia and Latin America is increasing (FAOSTAT, 2013). This is thought to be due to increasing fuel prices causing people to switch from motorised vehicles to equine-powered carts, alongside growing human populations. In rapidly industrialising countries, donkey populations remain relatively stable and are predicted to continue to do so as long as there are significant rural populations without access to motor transport. The global population of mules is thought to be rising, and the number of horses perhaps still static; in 2004, horse numbers were said to have been static for the preceding 30 years (Starkey and Starkey, 2004). In 2006, it was estimated that working animals supply approximately 50 per cent of agricultural power needs globally (Swann, 2006).

Roles of working equines

The majority of working animals are owned by individuals who use them as their sole means of income to support often large and extended families – between five and 20 people can be dependent on each animal for their daily survival (McKenna, 2007). Thus, in low GDP countries hundreds of millions of people depend on equines.

In addition to being used as pack and riding animals, equines, most often donkeys, are increasingly used for light cultivation tasks and drawing carts. In low-income countries, they are the most important source of agricultural energy and transport for resource-poor communities in both urban and rural areas (Swann, 2006). The use of donkeys as pack or draught animals has been key in enabling small-scale farmers to participate in the market economy. They transport everything from people, produce and building materials, to food, water and fuel. Donkeys are the key means by which the agriculture and food distribution systems in some nations function – animal transport enables small farmers to establish wider contacts with traders, improving access to markets and allowing them to increase production and profits.

Historically, men have tended to control animal power technologies, including ploughing and transport. Women, as major carriers of water, fuel wood, food grains and agricultural products can benefit particularly from transport animals. The donkey is said to perform the equivalent role for African women, of the washing machine for European women because the time saved using donkeys to carry water has enabled women to do other tasks and even to have some leisure time. For example, a study comparing two Maasai women fetching water, one using a donkey and one carrying it herself, indicated that using donkeys can save as much as 25 hours a week for other activities (Fernando and Starkey, 2004). In low GDP countries, working equines, particularly donkeys, can benefit all members of society, have a role in the empowerment of women and in easing the workload of children (Fernando and Starkey, 2004). Using donkeys in traditionally time-consuming and arduous domestic chores, such as fetching water and gathering firewood, provides women with more time and opportunities to earn extra money and to be involved in their community, both important aspects in raising the status of women.

The United Nations predicts that the population of Least Developed Countries will double from 804 million to 1.7 billion by 2050. Thus, the use of working equines is predicted to continue for many years to come.

Photo: Suzanne Rogers.

Welfare issues

Horses, donkeys and mules work alongside their owners for long hours, in difficult conditions but they are often overworked and poorly cared for. Colic, parasites, lesions, lameness and injury are common, as are various infectious diseases. The equipment (harnesses and carts) is often poorly made and designed, causing discomfort, lesions and injuries. The way that the animals are kept when they are not working rarely meets their needs: for example, tethering and hobbling without access to shade, food or drink is a common practice. Dehydration, malnutrition and exhaustion are common. Overloading is prevalent because the more that can be loaded onto a cart or animal directly or indirectly affects the amount earned for the job.

Few owners have access to veterinary advice or treatment. For example, in The Sudan, which in 2005 had the world's 25th highest equine population, there was one vet per 5.6 million animals (Swann, 2006), and in Ethiopia there is usually only one qualified government vet, with no training in equine medicine, in a district based at a clinic typically covering 40 villages (Swann 2006). As a result, many owners rely on traditional medicines and treatments, which are often inappropriate and cause welfare issues themselves. For example, firing (causing burns to the skin with hot irons or corrosive substances) is used in the belief that this will cure underlying problems and is common in the Middle East and Central America. And in Cambodia, monkey blood is used as a cure for colic, although this practice is decreasing due to the outreach efforts of the only welfare organisation in Cambodia concerned for working ponies.

Equines are a social species, but most working animals are often unable to interact with others. Many people who have worked with communities who rely on working

Table 9.1 Exploring the challenges to the Five Animal Welfare Needs for working equines.

Freedom	Comments
Need for a suitable environment	The environmental conditions animals work in are often extreme heat and cold with little opportunity for shelter. Heat stress is a significant problem for working animals in many countries.
Need for a suitable diet	Working equines are often underweight and provided with an inappropriate diet. Equines are often fed concentrated feed and given inadequate forage, meaning that not only are their nutritional needs not met but their behavioural needs are not met either. Working animals require 40–60 litres of water a day and are often given just 10–20 litres, therefore many equines are suffering from thirst and dehydration.
Need to be able to exhibit normal behaviour patterns	The way that working animals are managed often offers little opportunity for normal behaviour. For example, equines are social species but many are kept and worked in social isolation (although some may be in the proximity of others, they are often not allowed to interact). Abnormal behaviour, for example pica, the (non-accidental) ingestion of substances that are not food, for example cardboard and plastic, is also widespread. Most working animal projects focus on treatment of injury and disease because it is such a visible and pressing need in the field. There are few people that understand behaviour and notice abnormal behaviour but because this need is important for animal welfare, it is vital that projects advocate management changes that address behavioural needs.
Need to be housed with, or apart, from other animals	As described in the text, equines are social animals and need contact with other equines. However, if their history is such that they have not developed normal social behaviours, being in close proximity to others would be a negative experience. Therefore, the requirements for this need vary between individual animals.
Need to be protected from pain, suffering, injury and disease	Working animals suffer a range of chronic and acute diseases and health problems. One of the main causes of problems is working the animals at too young an age before they have developed fully. In some countries, the life expectancy of a working horse is as low as around 6 years of age (a horse could be productive into its twenties and a donkey into the thirties). Examples of health issues are: colic, lameness, broken limbs, skin infections, lesions, splints and spavins, laminitis, reproductive health problems and high parasitic loads, to list just a few. In many countries, local treatments are inappropriate and cause welfare issues themselves – e.g. firing (causing burns to the skin with hot irons or corrosive substances in mistaken belief that this will cure underlying problems) is common in the Middle East and Central America. Significant risks of injury from road traffic accidents and due to poorly constructed carts and stables.

animals, have observed that when donkeys are released from work to forage at the end of the working day, they gather together and interact with each other before seeking out water and drinking as a communal group. This suggests that socialisation is the first priority for fatigued and dehydrated animals, followed by drinking. At the end of the working day, when animals are allowed to free forage, their behaviour changes from appearing "depressed" to appearing like a grazing animal in other contexts.

As described in the Introduction, welfare is the mental and physical well-being of an animal.

Table 9.1 explores the Five Animal Welfare Needs in relation to working animals.

If we consider the welfare of working equines using the more modern "Five Domains" framework, it is clear that the quality of life is greatly compromised for arguably most equines who are "used" for work.

This welfare compromise is usually throughout their working life although subject to seasonal variation. For example, from the first day they are born, foals often follow the mares while they work and although it is good that they are with the mare for attachment reasons, this can cause physical development problems. Many working animals develop chronic long-term health problems such as lameness, lesions that never heal, dentition problems through poor diet, chronic parasite load and chronic problems associated with poor hoof trimming. Other issues, such as dehydration and pain from being hit during work, are acute and repetitive.

How can welfare be improved?

The needs of working equine and domestic equines are the same; it can just be more challenging to meet them. Or is it? In the western world we tend to consider some practices, such as hobbling and tethering, as old-fashioned and cruel, but if I were a horse and could choose between being tethered (according to good practice, where there was food and water available, a comfortable, appropriately designed tether, shelter, and so on) or stabled for 24/7 in social isolation with limited access to forage, then the decision would not be straightforward.

The root cause of poor animal welfare is arguably the way that humans care for and work their animals. Traditionally, animal protection organisations ran programmes providing treatment (veterinary or farriery) or aimed to "increase awareness" of issues that contribute to poor welfare. However, provision of veterinary care can lead to a culture of dependency, fail to reach all members of a community and have little or no effect on prevention. As we will learn in the next chapter, social science shows that awareness alone does not lead to behaviour change; nor does provision of knowledge alone – in the same way that we know that smoking is bad for you, or that we should eat five portions of fruit and vegetables a day, but people continue to smoke and not eat healthily, owners know they need to feed their horses properly, but if they don't have the money then can't do so. The saying *"Tell me and I forget, show me and I remember, involve me and I understand"* summarises one of the concepts behind comprehensive projects. Equine owners are not told what to do by outsiders, rather, through various activities, they own the project themselves and it is this crucial issue of *ownership* that creates involvement and ultimately results in changing the way the owners care for their animals.

Around 10 years ago, projects started to address this and move away from providing veterinary services, instead changing to community-based projects working with the people. The welfare of working equines in low-income countries is crucially important, not only for the health and well-being of those animals, but also for the livelihoods of those people dependent on them. Research in Ethiopia demonstrated that improving the health of donkeys means that they are then better able to work – healthier animals are more use than sick animals (Smith, 2004). Thus, the adoption of good equine health, welfare and working practices is among the most important ways that people in low-income countries can help secure and improve their earning potential.

Given the serious health concerns for working animals, it can be tempting to prioritise the health aspects of welfare rather than psychological elements. However, behaviour should be considered because there are ways to better meet the behavioural needs of working animals that do not require much investment in time or resources. We will now consider some examples.

Behaviour in mind: handling

I remember doing welfare assessments on a group of donkeys and horses in the Middle East. This involved following a standardised assessment protocol and recording the findings. The protocol included looking inside animals' mouths to estimate their age, approaching them to assess their degree of responsiveness, examining the underside of their tails (for parasites), and so on. The horses and donkeys in this area were very vigilant, nervous and fearful of humans, including their owners. To carry out the welfare assessment, I approached the first horse with behaviour in mind. I did not just stride up to the animal but approached slowly, pausing when they showed anxiety and continuing when they seemed "OK". By careful, considerate handling, vigorous scratches, fluid, calm movement and ensuring I never pushed the horse over her threshold, I was able to examine her completely. Just before my approach, I had used some hand sanitiser briefly, unaware of the intrigue this would cause. My every move had been watched because, unbeknown to me, this horse had a reputation of being very aggressive to strangers and impossible to examine. The local team I was working with had asked me to look at her first as a form of initiation. When she calmly let me examine her, the team was so surprised they started to form some theories about why this was and between the team and the owners, the sanitiser, which was not widely used in that region, became the favourite

Photo: Suzanne Rogers.

theory – it was obviously some special drug that gave me "magic hands", causing a calm horse. I became aware of much rapid talking and distraction and was told the theory; everyone wanted some of the special liquid. This presented an opportunity to explain that it was not the sanitiser that was responsible but an understanding of equine behaviour. A few horses later, a much-depleted bottle of sanitiser, the tests had been done to show that I might be right.

A common sight on a bustling street in areas where working equines are used is to see an owner pulling on the reins for opposing reasons – reins are often pulled and the owner shouts "yah" to start but also to stop and sometimes seemingly "just because". Understanding that the removal of pressure when the desired behaviour is given is likely to increase the likelihood of the animal doing that behaviour again (basic negative reinforcement, covered in Chapter 3) can change the way owners drive and ride their horses. A brief demonstration of well-timed application of negative reinforcement can provide the owner with a more effective way of communicating with their horse and because the horse understands what is required, beating is avoided. Given that equines are often driven for many hours in a day, this change could mean a significant improvement for the animals. Some projects swap sticks and whips for flags and rattle cans – eliciting forwards movement through fear of an alternative object. Such methods do not cause physical harm to the animal, as a whip might, but can still be used inappropriately so again, diligent training, using the principles of human behaviour change, which we will cover in the next chapter, is important.

Behavioural differences when in and out of harness

I once worked in an office where one of the staff had a beautiful golden Labrador guide dog. When his owner was working at his desk, the dog was mostly on the bed by the desk but often wandering around the office, raiding people's bins. When he was in his harness and out with his owner, he did everything you'd expect a guide dog to do – stopped at roads, didn't wrap his owner around lamp posts and certainly didn't try to stop and greet other dogs; he was in work mode. We were encouraged to take him for walks so that he had some variety when he went out walking with people who could see he didn't wear his harness but a normal dog collar and lead. The first time I took him out was a huge surprise – he pulled on the lead, wouldn't have stopped for roads without me holding him back, wrapped me around lamp posts and bounced about on the end of the lead like an exuberant puppy. On one memorable occasion, I let him off the lead and he bounced up to an office worker and in one practised action, swiped a deli sandwich out of her hands in mid-air on the way to delivering the sandwich to her mouth and gulped it down. I am not sure how convincing I was when explaining that I was walking the office guide dog, given that I was blatantly not vision-impaired and the guide dog wasn't acting remotely like a guide dog.

Guide dogs, you see, learn that they must put all their training into action when in the harness but that out of the harness, they have freedom to be a dog. The harness is what we call a discriminative stimulus – the dog is trained to learn that when wearing the harness, reinforcement is available and certain behaviours are expected and the dog can discriminate between when the harness is on and off. We see the same thing with horses – for example, many horses have learnt that when they are wearing a head collar there is no point in trying to "do anything" unless cued by the owner because unwanted movement will be punished. Working horses learn to suppress their behaviour when wearing a harness, when they can usually be relatively easily handled and examined. When the harness is removed, they are often a very different horse, showing much more fear behaviour and being more difficult to examine. Understanding this can help to keep us safe – it is counterintuitive to examine a working horse without taking off their equipment but with an understanding of behaviour, we know that if we do not take the harness off we are likely to be able to complete a more thorough examination. Often, it is possible to sequentially lift parts of the equipment so that from the animal's perspective, they are still wearing the harness.

Expect changes in behaviour

As a project takes shape, there is likely to be a change in behaviour of the animals involved and the project staff need to expect this. For example, I was involved with a project in Bogotá, Colombia. Here, horses were used in the recycling industry (they have since been banned from the city) – people collected various types of waste materials and took them to sorting areas where the waste was separated into different types, then taken to separate recycling facilities where it was sold for money. The sorting areas are where the people and horses lived and worked and were called "parking lots". When the project began, colic was very common and the horses were in poor condition, averaging 1.5 on a 0 to 5 body condition scoring system. The project was successful in changing the way the owners fed their animals, so successful that in just a few months the average body condition score had increased to 3! As the horses' health improved, their behaviour changed; they now had the energy to express their opinions about the way they were managed and handled and became what the owners described as "hot headed and feisty". When the horses were in poor condition, they were lethargic and unresponsive, but in better condition they were more difficult to harness up and handle. The owners had started to modify their equipment to make it "harsher" in an attempt towards compliance through pain (e.g. thinner nose bands and modified bits), which was moving towards creating more welfare problems. There is, therefore, a concern that in solving one welfare problem, another is created. Project staff were able to incorporate education for the owners regarding training and handling the horses and the project was back on track.

Considering equine vision

One of the main uses for working equines in some parts of the world is draught and pack work in buildings and structures such as mines and factories. In these premises, there are areas of darkness and light between the buildings or even in buildings in the case of mines. Considering that the horse's eyes take approximately 30 times longer to get used to changes in light than human eyes, means that when moving from a bright area to an area of dim light, the horse's eyes are not able to adjust as quickly as our eyes do and as a result they might be fearful of entering an area they are unable to see. If the handler knows this and gives a little more time so that the horse's vision can adjust, rather than beating the animal to enter, potentially dangerous situations can be avoided and the animal's welfare is not compromised through inhumane handling.

Training to accept harnesses and carts

In Cambodia, it is uncommon to see breeching as part of a harness. Breeching is the name given to the straps that form the "brakes" – preventing the cart from running into the back of the animal when going downhill or stopping. Instead, carts in Cambodia have padded fronts, and a very short gap between the cart and the animal; when the animal stops or goes downhill, the cart bumps into the animal but the impact is softened by the padding. This is not good practice and can result in severe physical issues including broken vertebrae in some cases. We wanted to introduce breeching and tweak the design of the carts and harnesses to avoid this problem. However, we knew that introducing such new designs is not easy; if people do not see the need to change, they are very unlikely to take up what you offer, and even if they do see the benefits, there are still barriers to adoption of new or alternative designs. I was also concerned about the behaviour element; the ponies would need to be taught about how the breeching works and introduced to the feel of the different harness in an appropriate way for the scheme to be successful. And it is this sort of thought process that illustrates how the lives of working equines could be improved with behaviour in mind.

Case study: Donkey Club

Anna Haines, an equine behaviourist who advises the Gambia Horse and Donkey Trust, brings the themes of this chapter together with this description of their donkey club scheme.

In remote villages in The Gambia, it is typical for the young boys of the family to be the primary caregivers for the working equines. The animals are sometimes handled poorly, and force may be used to move animals from one place to another due to a lack of knowledge about more compassionate methods. It is not uncommon for the animals to be left for long periods without food and water, often simply because the task has been overlooked by the young caregivers as they become occupied with other chores or playing with their friends. Anna explains how the Gambia Horse and Donkey Trust decided to address these challenges.

> In order to change human behaviour to create an improvement in the welfare of working animals, it is essential to establish what motivates the human caregivers.
>
> We developed a "Donkey Club" to create opportunities to engage with the young boys and develop positive relationships with the young equine caregivers. Through getting to know the children, we were able to establish that they were highly motivated by sports and enjoyed playing games with their peers. Donkeys were not typically seen as sentient beings by the children and were viewed as unfeeling modes of transport and farm machinery. It was recognised that for any long-term improvements to be seen in the welfare of working equines, these beliefs needed to be challenged and a programme was started to help the children to develop an increased level of empathy towards their animals.
>
> Children can attend Donkey Club with their own donkey and are encouraged to take part in team games with their donkey (Figure 9.2). These games involve rules to improve the ways that they communicate with their donkeys, through gentler handling techniques and more humane equipment, which was provided during the first session attended. Children are also encouraged to name their donkeys to help nurture the concept that they are sentient beings. Penalties are given to any children who were seen to be breaking the rules, and they have to sit out of the games for a certain amount of time for any inappropriate or rough handling. The motivation for

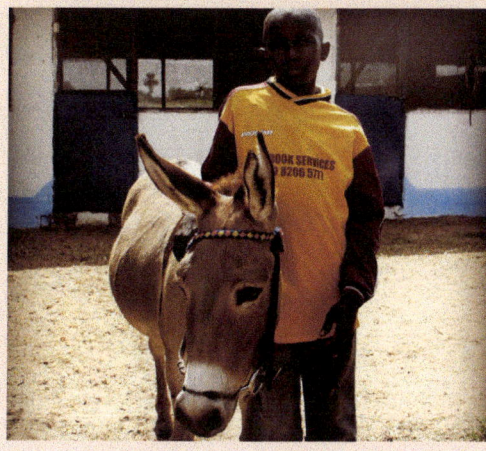

Figure 9.2: A boy and his donkey; Donkey Club members. Photo: GHDT.

the children to be able to take part in these friendly competitions is high, so these penalties are largely avoided. The Donkey Club sessions are also a great opportunity to help to identify any signs of sickness or injury in any of the donkeys and if any are noticed, this is used as an educational tool to explain to the children how to take good care of their donkeys, how to provide first aid to minor wounds and where to go to seek veterinary help if their donkey becomes unwell.

As the interest grew in Donkey Club, so did the number of children and donkeys attending. This led to the formation of village "teams" who enjoy competing against each other in friendly competitions, with the top four teams demonstrating their games in front of a large audience at the annual Horse and Donkey Show. This has become a real incentive over the years and has even been televised nationally throughout The Gambia.

Naturally, the donkey's welfare is a priority and the Donkey Club sessions are utilised as teaching opportunities for the young caregivers. Emphasis is placed on the importance of maintaining a strong, healthy donkey through the provision of good food, water, shelter, rest and veterinary care. Because the motivations of the children to be involved are high, they take these messages seriously and it has resulted in the Donkey Club donkeys becoming extremely well cared for. A mother of one of the young caregivers explained that her son was now taking his duties of looking after the family's working donkey very seriously. She explained how he had asked her to take care of the donkey whilst he was at school, but on one day she had forgotten to provide the donkey with any water. When her son returned home and found the donkey without water, he became very upset and said to his mother, "This donkey is an athlete and an athlete cannot perform without good food and water!" and he immediately went to the well to fetch the water for his donkey.

Through the process of developing a partnership with their donkeys over several Donkey Club sessions, it has also become apparent that the empathy that the children have for their animals gradually increases. This was really highlighted with one sad situation where a donkey named Blackie suffered a serious impaction colic. Ordinarily, these children would not have known how to get help for their sick donkey, but this Donkey Club attendee ran quickly to the next village to seek help from the nearest veterinary professional. Unfortunately, the colic was not possible to treat and the donkey passed away. The young caregiver was so devastated at the death of his donkey that he was inconsolable; this is an unusual situation in a culture where young boys are rarely seen crying, even in serious circumstances. The young lad was so devoted to his donkey that he refused to leave the body, and when the body of the donkey was taken away to be post-mortemed he insisted on staying with him throughout, whilst crying and stroking his fluffy little white face.

The first children to attend Donkey Club, approximately 10 years ago, have now grown into kind young adults who have kept their donkeys healthy and well cared for, for many years. They have become true ambassadors for the donkey and have helped to support the younger members of Donkey Club to develop the same knowledge and empathy for donkeys. Some of them are now working in the field of animal welfare or working towards a career in an animal-related field, helping to develop a brighter future for working equines.

> **Revisiting Samnang**

Samnang heard about a workshop taking place where representatives of communities across certain regions of Cambodia would gather to talk about working equines. He had always been interested in the health of his pony and, with the absence of equine vets in Cambodia, was seen as a source of information on sick animals in his community. As an influential and interested owner, he was invited to attend by the hosts – the Cambodia Pony Welfare Organisation (CPWO). They ran two workshops; the first was three days, the second five days with a month or so in between for them to practice their work in their community. The workshops did not involve any writing, which was a relief as he had never learnt how to write and focussed on discussions and participatory exercises to explore various themes regarding managing and working ponies. During the workshop, Samnang learnt about what ponies need, how to keep them healthy, what causes colic and how to prevent it, the benefits of managing resources by planning for seasonal changes, as well as that money is not the only barrier to improving things and there is much that can be done to benefit the ponies in his community and in turn the owners who rely on them for their livelihoods or day-to-day living. He learnt that the stable his pony was kept in could be easily modified to better meet the pony's needs; a mosquito net would help to prevent a common local disease, water should be provided all the time, not just in the morning, and that more space would allow the pony better quality rest. When he made these changes, other villagers saw and asked him about them; in just a few months, many owners throughout the community had redesigned their stables too and the changes were spreading. Samnang also saw his pony behave in a different way; he sometimes saw him sleeping and "dreaming" now that he could lie down more comfortably, he was easier to handle now that he knew not to wave his arms so much and it was nice seeing him interact with his other pony now that their stables were together. People were talking about the ponies in a different way and there was demand and excitement regarding visits by the CPWO vets who continued to coach, explain and facilitate change through gently supporting and encouraging the adoption of small management changes. Things were looking up for ponies in the region, and since the project was established in 2007 it has gone from strength to strength, improving the lives of thousands of animals across Cambodia.

Final thoughts

Projects that help working equines understandably often focus on the most pressing and visible needs. However, if a behaviourally-minded approach is taken, we can improve the everyday lives of horses by meeting their behavioural needs. This chapter provided some examples of behaviour-in-mind thinking and although focussed on working equines, much of what was covered could also be applied to equines used in sports or leisure, too.

References

Brooke (2015) *Invisible workers: the economic contributions of working donkeys, horses and mules to livelihoods.* Available at https://www.thebrooke.org/for-professionals/invisible-workers-research-project [accessed 26th February 2017].

Clarkson, N. (2017) World horse population likely to be over 60 million, figures suggest. Read more at https://www.horsetalk.co.nz/2017/07/10/world-horse-population-60-million/#OLfYx6tU1WhTWq5m.99 [accessed 21.02.2018].

FAOSTAT (2013) www.fao.org/faostat/en/#home

Fernando, P. and Starkey, P. (2004) Donkeys and Development: Socio-Economic Aspects of Donkey Use in Africa. In Starkey, P. and Fielding, D. (Eds). *Donkeys, People and Development. A resource book in the Animal Traction Network for Eastern and Southern Africa* (ATNESA). ACP-EU Technical Centre for Agricultural and Rural Cooperation (CTA), Wageningen, The Netherlands.

McKenna, C. (2007) *Bearing a Heavy Burden*. The Brooke, London. Available at www.thebrooke.org/__data/assets/pdf_file/0010/50968/BROOKE_heavy_burden.pdf

Smith, D. (2004) Final technical report R7350: *Use and Management of Donkeys by Poor Societies in Peri-urban areas of Ethiopia*. Centre for Tropical Veterinary Medicine, Roslin.

Starkey, P. and Starkey, M. (2004) Regional and World Trends in Donkey Populations. In Starkey, P. and Fielding, D. (Eds). *Donkeys, People and Development. A resource book in the Animal Traction Network for Eastern and Southern Africa* (ATNESA). ACP-EU Technical Centre for Agricultural and Rural Cooperation (CTA), Wageningen, The Netherlands.

Swann, W.J. (2006) Improving the welfare of working equine animals in developing countries. *Applied Animal Behaviour Science*, 100: 148–151.

UNDP (2016) Cambodia Human Development Report. http://hdr.undp.org/sites/all/themes/hdr_theme/country-notes/KHM.pdf

Chapter 10

Human behaviour in mind

Suzanne Rogers

Throughout this book, we have considered what a behaviourally-minded approach to keeping horses might look like. What we haven't yet considered, however, is how to make it a reality – how to change the cultural practices that are harmful to horses into practices that ensure that horses have a life worth living and are thriving in our care. And how to do this in a way that enables the people who love and "use" horses to be engaged. What we haven't yet considered fully is a behaviourally-minded way of changing *human* behaviour, and so we will turn our attention to this now.

Sam and Hardy

Sam is a strong, adventurous lady in her twenties who has grown up surrounded by horses. She was given her first pony when she was 4 years old: Penny, a chestnut Shetland with a penchant for breaking through any barrier the family could invent between her and the feed bins. They lived on her family farm and Sam's childhood was spent winning rosettes at pony club, having long horsey days enjoying being around each horse and pony as they came into her life before being sold as she "grew out of them". She competed at national level and might have gone on further had her interests not changed during her teenage years. Even through distractions she firmly remained horse-mad and through university and her first forays into employment she always kept a horse at livery for riding at the weekends and entering into the occasional competition. She loved the "horsey life" but had never really thought about the horse's needs beyond her received knowledge about how to care for them. The way her ponies and horses were kept at home and later, on livery, was traditional – some turnout, a lot of time stabled and the focus on having a neat, tidy bed, endless rugs, boots, potions and lotions. Sam considered her horses to live a pampered life. In her mid-twenties she bought a horse that she had seen online and couldn't resist. Hardy was a 16.2 Shire cross – a heavyweight horse with flowing mane, tail and feathers ... and an "attitude". Hardy did not like being stabled and showed his fear of confinement by jumping out of stables or kicking the walls through. He also did not respond to being "told off firmly" and quickly gained a reputation as being a dangerous horse. Under the insistence of the yard owner, Sam called me out as a behaviourist to see what I could suggest. Sam herself was very reluctant to seek such professional help as she considered herself to be an excellent horsewoman but also didn't want to be evicted from the yard. She described Hardy to me using some colourful language and the repeatable phrases she used in that initial call included "he's stubborn", "he has a bad attitude", "he needs to be taken down a peg or two", and so on.

During the time I spent with Sam and Hardy at the visit and follow-up, I observed their journey together. We'll come back to the outcomes later.

Considering change

Have you ever tried to change your behaviour? Most of us at some time have wanted to eat less, exercise more or stop smoking. Did you have the knowledge needed to make the change? Did you know what to eat, how to exercise, for example? Did you have the motivation to change? Did you understand the benefits of making the change? Most of us know that if we maintain a healthy weight we might avoid certain health issues, for example. And yet did you make the change and maintain it? If so, was it easy? If not, then why not?

We know from campaigns such as the "eat five portions of fruit or vegetables a day" that it is very easy for people to have a high awareness of the "rule" but not to actually live by it. We often understand the benefits of changing but despite having the desire, knowledge and motivation, actually changing our behaviour is difficult. And yet when it comes to animal welfare, we often assume that if we explain to people why change is needed they will instantly change the way they behave towards and manage their animals.

The root cause of compromised welfare in horses is human behaviour – what humans do, or don't do. When we ask horse owners to make changes to benefit their animal, we should be aware of the gap between knowledge/awareness and behaviour. As professionals helping horse owners, we tend to focus on short-term services and providing recommendations of how owners should address the issue. However, there is a mountain of evidence-based information about what does, and doesn't, lead to behaviour change in humans – it is found within fields including psychology, social economics, development approaches, behavioural change theories, counselling skills, social marketing, and so much more (Figure 10.1). As an equine behaviourist, and in other parts of my work as an animal welfare consultant, I started to recognise the need for us to apply what has been discovered from those fields to our work driving change for horses and I will outline some of the key principles in this chapter.

The four principles of human behaviour change

The output of the huge body of work exploring human behaviour can be roughly summarised into four principles: change is a process; understanding psychology is key in driving change; the environment influences change; and change must be owned.

Principle one: change is a process

What causes people to change their behaviour has been studied from various different angles and there are many useful theories to explore. For example, "Theory of Change" considers what is needed for change by starting with the goal, and identifying the precursors of each milestone – working backwards. It is a process used to create a strategy and can also be done retrospectively to understand how change happened. This approach is becoming increasingly used in strategic planning of

Figure 10.1: Some of the fields of research that are relevant to considering human behaviour change. Credit: Human Behaviour Change for Animals CIC.

animal welfare projects but can be a useful tool for behaviourists to use to take clients through the steps needed in complex cases.

The "Transtheoretical model of change" (Prochaska and DiClemente, 1983; Prochaska et al., 1992) outlines stages of change in individuals through five stages (pre-contemplation, contemplation, preparation, action and maintenance). The key message is that change is a process, it is not instant, and we need to understand where people are on that process to be able to support them in moving further along. For example, if a client does not truly appreciate the need to change and is not considering it, then if we suggest ways the change can be maintained we are not likely to be successful; the client required further motivation or information before they could consider preparing for the change and actually doing the new behaviour. Other models depict change as less linear with feedback loops and cycles. This concept of change as a process is useful for people working with individual clients such as vets and behaviourists.

Sometimes it is important to remember the bridges. When we want to change human behaviour, it is often tempting to provide the solution (the other side of the bridge), and explain why it is so important to get there for the well-being of animals and for ourselves. We motivate the person or community to want to be there and help them to see what it would be like once they are there. But if we forget to show them the bridge, how to get there from where they are now, and that it is safe to cross, they will stay in the security of where they are now.

Alongside the need to "create" or change behaviour is the important element of embedding it, so that the person does not return to their previous ways. Jo White, co-founder of Human Behaviour Change for Animals CIC explains:

> Research into the formation of habits suggests that it presents a powerful tool in creating positive behaviours, together with understanding and addressing negative ones. This is particularly relevant for activities undertaken routinely, which is pertinent for those engaged in looking after horses. Whether it is feeding to basic hoof or health care, habit formation could be the key to delivering sustained positive management changes.

In practice, as behaviourists, vets, and fellow horse owners, we need to support people through the process of creating good habits. We need to be innovative and steadfast in this stage because otherwise our work in the stages up to this point will be wasted.

A tool has been developed that helps users to understand and plan their behaviour change project – the COM-B model and behaviour change wheel (Michie *et al.*, 2014). The behaviour change wheel brings together theory-based tools developed in behavioural science to understand and change behaviour; it facilitates a step-by-step intervention guide. The first steps are especially useful in practice with horse owners to consider the problem, define the target behaviour, identify what needs to change and how to bring about that change, and I have applied this framework of thinking both to better understand the horse's behaviour from the horse's point of view and also the owner's target behaviour.

The main elements considered in the overall process of change are the triggers for behaviour change, the connections between the points and whether they can be mapped to better facilitate change. These elements can be applied at all levels from individuals to mass behaviour change.

Principle two: understanding psychology is key in driving change

This principle explores areas including: how much change is someone's autonomous decision and how much is a result of influence by others; how the mind works in processing new information; what factors affect our motivation for change; how barriers for change are often very deep-seated beliefs and values and how to best address this; and much more. If we have a better understanding of what motivates and influences people, we can apply the knowledge directly to our work.

Understanding the motivation for change and how new behaviours are deserted or maintained is necessary in planning effective projects or driving change. An understanding of the relationship between behaviour change of individuals and how that translates to increasing the dissemination of information and change throughout a community, is vital in planning and adapting projects that rely on the spread of best practices.

I will introduce four key concepts in this section: confirmation bias; confrontational versus empathetic communication; transactional analysis; and the righting reflex. Confirmation bias is the phrase used to describe the understanding that we are more likely to take on information that fits with what we already know and believe than information that challenges us. If we believe that horses thrive when confined for 24 hours a day then we are more likely to notice "messages" that back up that view – for example, messages from the marketing departments of products related to keeping horses in stables – and we are less likely to notice messages suggesting that horses do not like to be confined. To counter this effect, we can focus on common values: if we create dialogue focussed on the similarities between us and our "target" by focussing on the values we share (for example, a love of horses and a desire to solve the problem), we have a good foundation for more in-depth exploration of our differences. We can see an example of this when we revisit Sam and Hardy at the end of this chapter.

In terms of the psychology of communication, studies in the field of motivational interviewing (a counselling approach that is aimed at eliciting behavioural change) show that confrontation in a conversation is the most robust predictor for failure of the client to change whereas empathetic communication is a strong predictor of change. The challenge for behaviour professionals working with clients is the need to impart an understanding that the client's behaviour is having a negative effect on the horse without a confrontational element and, sadly, few behaviourists, vets and other professionals are trained in such counselling skills.

Transactional analysis is an accessible model that is useful in tracking your counselling performance during interactions with clients. It is based on the premise that we all can be communicating from one of three "states" – the parent, adult or child. In the parent state, we might overly impart information, in a way that is perhaps a little condescending or top-down. This might create feelings in the other person causing them to react from a "child state" – the person might feel inferior, disempowered and might experience some negative feelings associated with school and education. The "adult state" is where we are operating with logic, reason and level-headedness at the forefront. In any interaction, each person can be in any of the three states but the aim is to maintain adult-to-adult communication in a consultation setting. In practice, if we feel that we are teaching too much, or that we are ourselves being nudged into our child state by a client, and can recognise it, we can consciously get back to adult state, for the most effective communication.

The righting reflex is another motivational interviewing term that describes the human desire to provide solutions for others. However, when we make suggestions and provide ideas, counter-intuitively it is likely to have the opposite of the intended effect – when faced with one side of an argument we are likely to bring up the other side, so you might generate more ideas against your solutions than you do support for them. Also, by giving the client an opportunity to vocalise reasons why not to change, for example, you enable those ideas to become more strongly held, or more embedded. Providing solutions can therefore disempower your client when you are aiming for the opposite result. Instead of providing solutions we must develop skills that help people to come up with their own solutions and ideas, where we are the facilitators.

Another model that illustrates the psychology of change in a nutshell is the "elephant and rider" model coined by the Heath brothers (Heath and Heath, 2010). Imagine a rider on an elephant (which is not condoned as elephants are wild animals and the process of training them to accept riders is never kind; see Figure 10.2). The rider represents the rational, logical side of how we make decisions to change and the elephant represents the emotional side of our decision-making. The elephant is much larger than the rider, just as our emotions have a much larger role in our decision-making than we might appreciate. Therefore, to motivate, inspire people and change behaviour we need to reach the emotional side of our audience. Again, this reiterates the importance of truly considering how we communicate with clients and not just the content of our message.

Figure 10.2: Elephant and rider model. Illustration: Kirstin Calvin.

As professionals, behaviourists, vets and trainers are teachers, educators, facilitators, counsellors, persuaders and much more but are often not trained in those skills. I cannot stress enough that to change the world for horses we need to develop those skills and value their importance just as much as we value our other professional knowledge and expertise.

Principle three: the environment influences change

Continuing with the elephant and rider analogy: if we just have a person riding an elephant without a clear path they will just be meandering through a forest. The third element needed for change is a clear path; in real terms this could mean a strategy or the right environment in terms of education or legislation – if we are not acting with a supportive legislative background we will be more limited. Also, barriers to progress are considered in this principle.

How do you facilitate change, break down barriers, create social trends and encourage new dialogue that becomes the norm? Social marketing is the main discipline to look to for the answers. Well-used in the health and environmental sectors, social marketing identifies barriers to change, proposes solutions, works to enable change by providing a suitable environment and uses concepts from group psychology to drive social change. Social marketing is mostly relevant for "mass" change but can also be used on a smaller scale to encourage take-up and spread of ideas through communities, including the equestrian community.

The environment is a significant factor maintaining some of the negative practice in the equestrian sector. Livery yards with their rules and cultures surrounding how horses are managed, disempower individual owners and create an environment that is not conducive to change. Routines become embedded and behaviour is often not at the heart of management decisions. There are many barriers to improving the way horses in a livery yard are cared for, but change can happen. I have worked with one yard who redesigned their fields to better cater for the animals in their care, changing the position of water troughs to enable social drinking and removing fencing that created narrow areas where horses could become "trapped". Another yard owner, who saw the difference having a pen outside one stable made for a horse who was particularly struggling with confinement, created pens outside every single stable on the premises, doubling the space available for each horse. And one riding school I worked with changed their management so that enrichment wasn't an added luxury but an integral part of the daily routine.

Education also falls under this principle as it is another way of providing the "path" in our elephant and rider analogy. Any discussion considering the reasons behind animal welfare issues will identify a lack of education as one of the root causes, yet education so often is not given the attention it deserves. Jo White explains:

> At HBCA, we are passionate about education, as it presents the opportunity to proactively address the need for positive human behaviour development and change as a preventative measure for avoiding animal suffering. If we can instil compassionate behaviours in children, we will not have to re-educate the adults! As the future generation, children drive cultural change, so they are the key to sustainable change.

The element of "culture" and learnt behaviours is hugely significant in the horse world. Indeed, those of us who have been involved with horses for most of our lives have so much received wisdom that sometimes it takes a non-horsey friend to question what we do for us to even notice our embedded beliefs and behaviours! But culture and learnt behaviour is no excuse for cruelty, and culture and knowledge can change, sometimes very quickly.

Principle four: change must be "owned"

There is a saying, "Tell me and I forget, show me and I remember, involve me and I truly understand", which perfectly illustrates this principle of change. People need to truly appreciate the relevance of the desired behaviour change to them for change to happen. If we understand that people learn and change if they are not just told what to do through resources or typical top-down educational outreach, not just shown what to do through demonstration, but are truly involved in the process of change, we can facilitate that change. This process involves enabling people and communities to explore issues and come up with solutions themselves rather than "train them" to implement a preconceived solution.

People only change if they believe they can; this is the essence of self-efficacy theories. Self-efficacy can be thought of as a task-specific version of self-esteem. Individuals are more likely to engage in activities for which they have high self-efficacy and less likely to engage in those for which they do not. In practical terms, this again highlights the need for us to provide a safe space for owners to try out new behaviours, under our guidance, and not just tell them what to do and then leave them to try it alone.

The concept of positive deviance also comes under this principle – an approach based on the observation that even though most individuals or groups in a community usually have access to the same resources or face similar challenges, some find better solutions than others. Usually considered to be a community-driven approach, it enables people to discover these successful behaviours in their communities and develop a plan of action of dissemination. However, we can also apply positive deviance to our work with owners and even ourselves. If we look for the good things, the good behaviours, the small successes, highlight them and build on them, we can drive change.

These principles are only the tip of the iceberg in the vast amount of research exploring the fascinating field of human behaviour change. Concepts do not always neatly fit into one principle or another but reflect our current framing of the concepts that underpin what is included in the study of human behaviour change; there are many alternative ways they could be categorised.

Problem-solving with behaviour in mind

Let's consider how having equine and human behaviour in mind is used by a behaviourist. Let's first consider the background before we make the first visit to the client. When people encounter problems with their horses they tend to go through a range of responses – some immediately call the vet as they suspect, or want to rule out, pain as a reason for the behaviour. Others might change the horse's diet or start adding a supplement to their feed; others might turn to a specific training or horsemanship method, get the saddle checked, turn to a herbalist or aromatherapist, some might wait and see if the behaviour gets worse or disappears without any intervention, and so on. With so many professionals in the horse industry, and so much information and "communities" on line, there is no shortage of people to turn to – and this brings both positive and negative effects for horses. In terms of the process of change, if the horse's changed behaviour is considered the goal, when an owner has recognised that they need help with their horse, they move through the stages of pre-contemplation and contemplation as they decide to engage some help. However, they might fall back through those stages due to peer pressure from other people on the yard, and so on. As a behaviourist, it is useful to fully understand the person's motivation for calling you out so you can start to explore their values, which is a key part of building rapport with clients and the empathic conversation needed for change.

As described in Chapter 8, a high percentage of behaviour problems are rooted in present or past pain. When pain is addressed or ruled out as a cause, sometimes the unwanted behaviour goes away but sometimes the horse needs help to relearn. Many trainers and owners turn to "methods" of training for the answer. As we saw in Chapter 3, some seemingly kind approaches are not so kind when we consider learning theory. In addition, although training might help in some situations, it often doesn't address the cause of a problem but rather addresses the symptom. For example, let's consider a horse who is bucking when ridden. The owner first rules out pain, by having vet and saddle checks, and then turns to a trainer for help; who, depending on what method they advocate, might suggest several courses of action. A riding instructor might suggest schooling; a natural horsemanship proponent might advise that the horse needs more groundwork in an attempt to improve the relationship between horse and owner on the ground before resuming riding; a clicker trainer might suggest the use of reward-based training to improve the relationship and ridden work and to establish a positive association with tack; a herbalist might suggest a calming supplement. All these approaches might be effective to a greater or lesser degree. However, the one thing they have in common is applying a tool to tackle how the problem manifests, not the cause of the problem. As such, only looking at a problem through a restricted lens might overlook important issues and could make the problem worse or put horse and owner in a dangerous situation. Again, communication is key; the owner might well have turned to solutions you don't think are appropriate and how to navigate this requires skills following human behaviour change principles.

Only by addressing the root cause of the equine issue, and any limiting beliefs or embedded behaviours on the human side, can we be sure to solve the problem effectively and safely and ensure that we won't just get another symptom emerging as the problem is being tackled. Also, it wouldn't be ethical to train a horse to put up with pain or fear (even if you use reward-based training) without tackling that underlying pain or fear; and it wouldn't be ethical to focus on training the horse if the rest of their life is so "unhappy" due to their needs not being met that they aren't in a position to learn. Many riding instructors, back specialists, trainers, nutritionists and other professionals would consider the management/whole picture but these are few and far between. This book has explored what a "behaviourally minded" approach across equestrianism would be like, and in this chapter we have started to explore the importance of the owner's behaviour so that we can ensure the necessary changes are made.

Finding the cause: the process

In a consultation, the behaviourist would first take a full history. They will ask lots of questions and some might seem irrelevant at the time, but the behaviourist will be building up a picture of the owner and horse – their partnership, the owner's experience, attitudes and aspirations, the horse's background and previous experience that might be relevant to consider later, the management regime, what has been done to solve the problem so far, any welfare issues that need to be discussed and anything that highlights the need to involve another professional, such as a vet or nutritionist. The behaviourist will then provide his/her thoughts on the main elements of the problem and start to talk through the approaches to solve it. There is likely to also be an element of observing the horse and environment and perhaps on training or handling, although this is often in a later session depending on the problem. The behaviourist will work with the owner to put together a behaviour modification programme – it is no use imposing a plan on an owner when they don't have the time or inclination to carry out the recommendations. It is vital that the behaviourist acknowledges this and ensures that their suggestions are practical and supported by the owner, using the key principles of human behaviour change. Consultations aren't like TV programmes – with an aggressive horse, or one that bucks when riding, the behaviourist is unlikely to suggest that the horse is put in the situation where they will show that behaviour as this would not be safe. It is also unnecessary if the behaviourist is skilled at questioning.

The behaviourist will consider all 24 hours in the horse's day, to determine how well their needs are met, as covered at length throughout this book. So, what about problems that remain after the management has been addressed or in horses who live out in social groups in an enriched environment? Well, then behaviourists go through all the possible reasons for the development of the problem behaviour and suggest approaches to solve it. There are five main elements to this process:

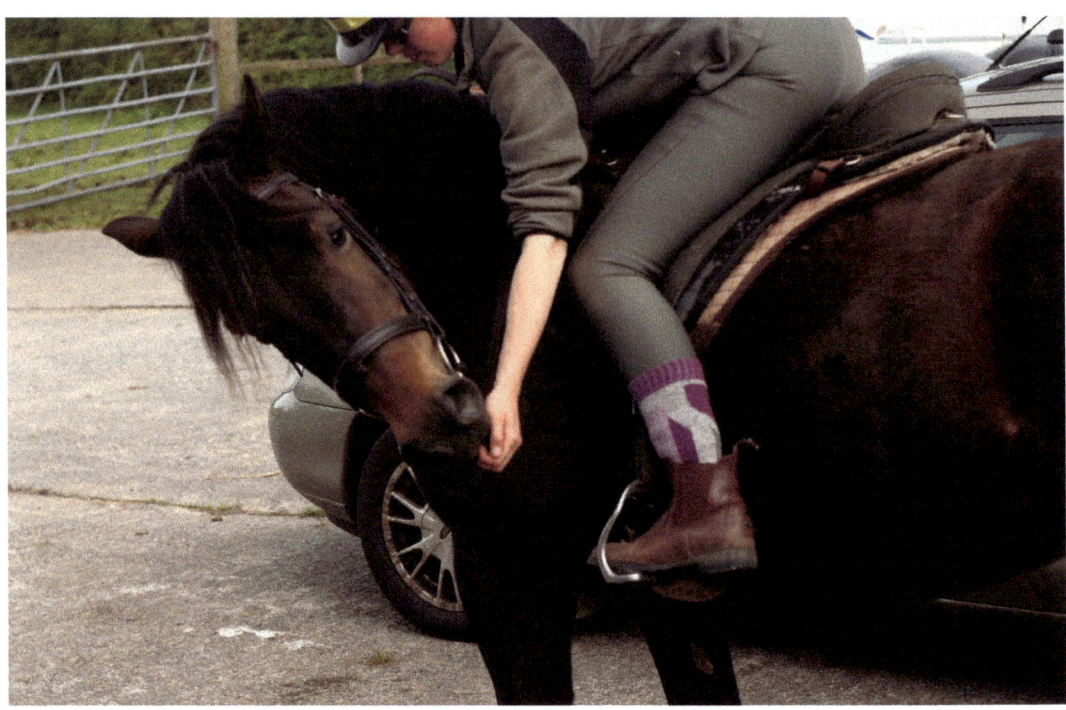

Photo: Anthony Payne.

1. Considering whether the behaviour is normal for horses, normal but out of context or abnormal. Many behaviours are normal for horses but are unwanted! Eating a small amount of bark is normal for a horse but if your horse is eating your prize apple trees then this is unwanted behaviour. The behaviour might be normal but out of context – to eat a small amount of bark/wood is normal but for a horse to eat his/her way through a stable door is not! Or the behaviour might be abnormal; eating non-nutritious substances such as plastic or sand (known as pica) is abnormal. It is important to understand which category the behaviour falls into before considering how to modify, provide an alternative outlet or prevent it, and behaviourists use their extensive knowledge of equine behaviour to diagnose the problem.
2. What learning is involved? As introduced in Chapter 3, there are many different ways that horses learn, and behaviourists have a full understanding of this which is important when considering how to train them to do something different. If a behaviour has become automatic (classically conditioned) then it will need to be tackled in a different way to if the horse is still learning about an object or experience. The animal developing a phobia of something is different to the horse just being anxious about something.
3. An understanding of physiology is important. In Chapter 8, we learnt about the physiology of stress and the impact on behaviour. Another example is that the chemicals involved in the biology of **aggression** mean that movement can make it worse, thus it is important to do slow or stationary work with horses with **aggressive tendencies**. There is certain physiology associated with stereotypical horses that make these behaviours addictive and many sexual behaviours have physiological

aspects that must be considered in order to provide an appropriate plan of action.
4. Welfare: behaviourists care about the animal and the human, and safety and welfare is of paramount importance. If a horse is suffering through management or training regimes, the behaviourist will work with the owner to address this as a priority.
5. Owner–horse relationship: behaviourists also have counselling skills, as often part of the problem lies with the owner's perception of the problem, confidence in themselves, own fears and concerns and expectations.

These five elements are put together to construct a plan of action for the owner, with help and support from the behaviourist, to work through with their horse to address the problems. This is likely to involve management changes, handling and training practice, and of course input from vets or other professionals as appropriate. But, crucial to all those things, the behaviourist will apply the principles of human behaviour change to every stage and every interaction with the horse owner to drive change and embed new behaviours.

Considering wider change

Often behaviourists, vets or other equine professionals observe situations where equine welfare is compromised and want to do something about it, whether with an individual owner or at the community or even policy level. So how can we help more horses, not just the ones we come across in our day-to-day professional lives? There are many groups of people planning and implementing campaigns and actions on the basis of what "feels" right to do and what should make a difference, but how could this be done in a more strategic way?

There are many different types of possible interventions to change behaviour; each has advantages and disadvantages and must be considered on a case-by-case basis and in the wider context. In this section, I will list some typical ways we can drive change and discuss each briefly.

Legislation and enforcement

A common reaction to address a welfare issue is to attempt to ban the practice that caused it. However, without a full understanding of the country-specific legislative process and enforcement procedures it is easy to spend time campaigning for changes that are not needed due to existing legislature or local by-laws, would never be adopted, might result in different welfare issues or are simply inappropriate for the situation. For example, the practice of tethering horses (tying them to a stake in the ground so that they are restricted) is common worldwide but there are welfare concerns including: horses can be injured from the ropes or restraining materials used; the animals can get caught in the equipment causing further movement restriction, discomfort or injury; the animals can eat all the grass in the area and then not have access to appropriate forage; the horses can be at risk from loose dogs or cruelty from people if tethered in a public place; and animals are sometimes tethered without access to shelter or shade. Activities towards obtaining a legal ban on tethering might seem a logical approach. However, let's consider this with behaviour in mind. If this management system was taken away, what would happen to the horses? Would they instead be likely to be kept in small huts or warehouses, away from the public eye but in compromised conditions?

In some cases, legislation exists already that would effectively ban the animals from being tethered in this way. For example, in the UK there is the Control of Horses Act 2015, which bans grazing on public land: given that horses are generally tethered on public land, tethering is effectively banned but the law is rarely enforced. Likewise, the Animal Welfare Act 2006 safeguards against lack of food, water, shelter, injuries and pain so some of the welfare concerns regarding tethering are already illegal with proven legislation. All these issues must be considered before lobbying authorities for legislative change. I believe that much energy is wasted in trying to introduce bans that are ill-thought-out, already covered through other legislation or that would have concerning implications.

Perhaps the most significant issue in terms of legislation is enforcement. In most countries, there is some legislation to protect animals, but it is often subjective, and without test cases and common usage, prosecution is unlikely to take place. In cases where the instinctive reaction is to get a "ban", it would be pertinent to fully consider all the legislation currently in place, and research what entity is responsible for enforcement, what warning systems are in place and previous records of prosecutions using that legislation, before deciding on the most appropriate strategy. Perhaps an awareness campaign regarding existing legislation might be a better approach or perhaps enforcers need training in how to recognise compromised welfare, and so on.

Developing animal-protection legislation is a valid approach and can indicate progression towards a more compassionate society. For legislation to be effective, the resources must be available for enforcement and the societal background must be conducive to compliance with the proposed legislation.

Regulation and control

Regulation schemes can help to bring different service providers, livery yard or riding school owners, for example, up to certain standards, reward compliance, enforce rules, and therefore drive incremental progress. Membership schemes, where membership is dependent on meeting certain standards and is subject to renewal every 12 months, can also help to inspire members to improve standards, create a cohesive community of service providers or owners and be financially self-sustainable. However, the main concern is that by introducing minimum standards through registration schemes, the minimum then becomes the "goal" and progress can stagnate. Also, if the scheme does not include the majority of service providers there might not be much incentive for people to join in. I can dream about a "behaviour-friendly livery yard" network that genuinely meets the needs of horses ...

Membership schemes utilise some key principles of the science of human behaviour change. For example, by "labelling" themselves as members of a scheme or livery yard that stands for high welfare standards, people are then more likely to act in accordance with that label. For example, if one of the criteria for membership was that the owner's horse has daily access to time in a field or area for free movement, then through being "labelled" as a member, people are more likely to continue those behaviours of a good owner with which they now identify.

Social norms are also relevant here – norms are the rules of behaviour that are considered acceptable in a group or society. People who do not follow these norms might be shunned by other members of the community or suffer another consequence. Norms can change – for example, being part of a membership scheme could become a norm, which would provide motivation to be part of it and therefore welfare-friendly management.

One of the key elements of successful regulation and membership schemes is that they are created with input from the community itself; the scheme will then be locally relevant and there will be more "buy-in" to the standards set than if a scheme was imposed by authorities or outsiders. Another key element needed for success is good management of the logistics – inspections, membership databases, and so on – which can be a significant drain on limited resources.

Incentive-based schemes

Incentives can be positive or negative: charging fines for breaching certain standards is an example of negative incentives; giving rewards for exceeding standards is an example of positive incentives. In the context of equines in competition settings, negative incentive schemes are common in the form of inspections, fines or disqualification but, in my opinion, behavioural elements are not taken into account with any real meaning. Inspection systems can be run by authorities: at regular intervals, owners of equines used in tourism present for inspection and are fined if their horses or equipment do not meet certain requirements. The fines can be used to fund the Inspectors, rendering schemes financially sustainable. The challenges associated with such schemes are usually based in the lack of involvement of the owners: the top-down inspection approach can generate negative feelings towards the authorities and the owners are less likely to be enthused to make changes or to strive to meet higher standards than the ones required to pass an inspection.

Community engagement

Interventions that focus on empowering the community to drive change from within, perhaps through education, skills development, and so on, can help to address key welfare concerns and it is in this area that the author's key interest lies. A community could be defined geographically or be interest-based (e.g. an online community of people interested in a certain breed of horse). However, there are certain factors that must be properly considered for community engagement to be successful. For example, if the community does not perceive that change is needed there is unlikely to be any enthusiasm to work towards such changes imposed by an external organisation. Likewise, providing education might not address welfare concerns if lack of knowledge was not the barrier to good practices. Usually there is an element of knowledge needed but the issue of access to resources is much more of a barrier to the desired behaviour.

Community-based interventions can struggle to become sustainable as the intervening organisations often lack the funds and ability to support communities through change, to help them address challenges encountered along the way and to address issues in the wider context. Even when change does happen, therefore, practices sometimes revert back after the intervening organisation leaves (remember the Transtheoretical Model of Change introduced earlier – behaviours can slip back at any point along the change process).

Successful approaches truly involve the community in exploring the current situation and the challenges they face, and then planning solutions takes place by the community, for the community. This approach can lead to sustained change as the communities take responsibility for the welfare of the animals in their care and generate strategies for solving issues that arise.

On a good day, I am optimistic that the community of horse lovers can change so that horses are not negatively affected by people wanting to ride and spend time with them. However, on a bad day, the embedded psychology and culture of the way we use horses seems insurmountable. The source of genuine optimism is that the community of people who do recognise the effect humans have on horses is growing, rapidly.

Alternative technologies and livelihoods

Interventions that attempt to reduce or stop the use of animals by introducing alternative technologies illustrate another approach. For example, interventions to replace horse-drawn vehicles with motorised vehicles might be successful in some communities. Sometimes, such approaches might have negative effects on the community as a whole – for example, increasing the demand for fuel might not match the supply.

Encouraging a horse owner to give up their business hiring their horses to tourists for riding and instead to adopt an alternative livelihood might be successful in some circumstances but is likely to have many challenges. For many service providers, their activities are more than a livelihood but part of their identity. Well-run schemes are fully participatory and explore the person's inspirations and aspirations before considering the locally available livelihood options and eventually supporting the person in the adoption of a new livelihood until they are established. With respect to the leisure horse, some owners would arguably be more suited to owning a rocking horse or horse-riding simulator, but it is unlikely that they would embrace that suggestion!

Raising awareness

A typical approach is to raise awareness of a welfare issue to drive change and this can be successful if that awareness leads to behaviour change. "Intelligent worming" is an example – the information that instead of worming horses according to a calendar, whether or not they needed it, samples can be taken, sent off and the animals only wormed if required, has become increasingly popular. This method of worming is better for the horses, as it limits the chemicals in their bodies but also for population health, as resistance to wormers will not develop as fast using this approach. For many owners, awareness that this approach is available was enough to cause them to adopt the system.

However, for other issues this approach will not be successful; the reasons depend on the issue, but one example is that if owners feel an awareness campaign is criticising their sport or management, the outcome might be that the criticised behaviours become even more firmly embedded. Raising awareness can be successful when the unwanted practice is already recognised as something that should be reduced – in these cases, drawing attention to the practice could be beneficial and lead to sustained change.

In summary, to plan an effective campaign or to drive change requires first that the situation is well-researched. Only when we understand the true causes of the issues, the wider context, the stakeholders involved and the factors responsible for maintaining the behaviour, can we plan how to change it.

Sam and Hardy

We left Sam and Hardy struggling, but two years later so much had changed. Sam had totally changed her attitudes, knowledge and behaviours regarding not only Hardy but horses in general. Her reasons for owning a horse had completely changed and now she owns a herd of happy, enriched horses, not all of which she rides. Why did she change?

In the first consultation, it was very clear that Sam had very strong opinions about what horses should and shouldn't do, and why Hardy was so difficult to handle. Those opinions were not founded on the evidence base regarding animal welfare, learning theory, ethology, and so on. During the history-taking stage, I not only had to be careful not to condone what she was describing or be seen to agree with her reasoning, but also it was important to give her the time to explain the situation. Through the use of reflective questioning, I gained a clear understanding of the issues and also identified some of Sam's core values and the belief frameworks she was operating under. This enabled me to explain and summarise the situation using language that matched her values, minimising the risk of confrontation yet staying true to what I needed to explore. Using all my training in communicating through challenging situations, we were able to put together a plan of actions to try. During the consultation, I demonstrated some of the handling approaches we needed to introduce and set up opportunities for Sam to rehearse the new behaviours in a safe environment, trying to create self-efficacy regarding key activities that would need to be done daily. Through the language I used, guided observations and progress seen, we could gently explore the true nature of horses as sentient beings, with their individual personality traits. And in time, Sam's attitudes changed from considering horses in a very utilitarian way, enjoying what she could "make them do" more than for their own intrinsic value, to the point where she no longer wanted to do any activity that was not in their best interests. She stopped riding the individuals that seemed to struggle with various aspects, worked through issues with others using established techniques and stopped riding Hardy altogether as she felt he was much happier exploring the woodland in hand rather than when being ridden.

Final thoughts

This chapter has outlined some of the ways that behaviourally-minded approaches could be adopted by equestrian society. It has introduced the subject of human behaviour change, as a multidisciplinary approach to driving change and improving the lives of animals.

The whole book has provided an insight into the concept that having "behaviour in mind" can help incrementally improve the lives of horses whatever they are "used" for and is optimistic that if this way of thinking were to be adopted, then many people would reconsider what we expect of our horses in the first place.

References

Heath, C. and Heath, D. (2010) *Switch: How to change things when change is hard*. Random House Business, New York, USA.

Michie, S., Atkins, L. and West, R. (2014) *The Behaviour Change Wheel*. Silverback Publishing, Surrey, UK.

Prochaska, J.O. and DiClemente, C.C. (1983) Stages and processes of self-change of smoking: toward an integrative model of change. *Journal of Consulting and Clinical Psychology*, 51: 390–395.

Prochaska, J.O., DiClemente, C.C. and Norcross, J.C. (1992) In search of how people change: applications to addictive behavior. *American Psychologist*, 47: 1102–1114.

Index

abnormal behaviour
  biting/self-mutilation 35–6
  neurobiology 27–8
  prevention 27
  stereotypies
    box-walking 26, 27
    coping strategies 168–9
    cribbing 26, 27–8, 48
    high incidence 26–7
    lip smacking 26
    maladaptive elements 168–9
    misinterpretation 26
    repetitive head nodding 26
    tongue playing/lolling 26
    weaving 26, 27
    windsucking 26

Bedouin-style trail rides 94–6
behaviour
  aggressive 54, 151, 170
  avoiding unwanted behaviour 125, 132, 133
  fear/confusion 55
  needs
    association 8
    body care 7
    eating/drinking 7
    exploration/investigation 8
    motion 8
    rest/sleep 8
    safety 7
    use of space 8
  nutritional effects
    concentrates 24–5
    evolution of the horse 24
    feral horses 24
    starch 25–6
  problems 6
  stereotypical and confinement 26–8
BHS *see* British Horse Society
breeding 33
  adolescents 49–50
  artificial insemination 40–1
  early training 45–6
  embryo transfer (ET) 40
  foaling/new arrival 42–3
  free-ranging conditions 37
  hand-breeding 40
  man-made restrictions/interventions 39
  mares/stallions in sport 40–1
  orphans 46–7
  pasture breeding 40
  pregnancy 41–2
  stallion behaviour 35–8
  weaning 47–9
British Horse Society (BHS) 117
Bureau of Land Management Burros 138

Cambodia Pony Welfare Organisation (CPWO) 195
case studies/examples 2
  behavioural issues 53
  breeding 32, 35, 38, 45
  elderly horses 99
  equestrianism 79–80
  human behaviour 197–8, 210–11
  making changes 3–4
  pain/behaviour link 162–3
  positive reinforcement training 70–1
  rescue/rehabilitation 137, 146–7, 157
  stabling/exercise 12, 29–30
  weaning 49
  working animals 184–5, 193–4
CDS *see* cognitive dysfunction syndrome
CFH *see* Communities For Horses
change process *see* human behaviour change
'chase and charge' game 88–9
chemical restraints 174–6
  intramuscular 175
  intravenous 175
  oral 175
  sedation drugs 176
cobweb analysis 5

cognitive dysfunction syndrome (CDS) 105
Communities For Horses (CFH) 73
competition treatments
    bathing 83–4
    clipping 84
    pulling a tail 83
    solariums/hydrotherapy 84
    thinning manes 83
Concours d'Elegance 92
confinement 177
    companionship 177–8
    cross ties equine 179
    forage/browsing 178
    introducing movement after 179–80
        station-focussed walking 180
        target-stick walking 180–1
    making choices/exploring 178–9
    sleep 178
Cooke, Paula 81
counter-conditioning 63, 171–2
CPWO *see* Cambodia Pony Welfare Organisation
cross-country riding 87–8

diseases
    arthritis 103–4
    brain tumours 164
    Cushing's disease (pituitary pars intermedia dysfunction, PPID) 106, 163–4
    dysfunctional thyroid 164
    granulosa cell tumours 164
    petit mal/absent seizure epilepsy 164
    retained testicular tissue/undescended testicles 164
    see also health
Donkey Sanctuary (University of California, Davis) 138
donkeys 138–40, 186–7
dressage 84–6
    FEI rules 116
driving 92–3
drugs 25
Dujardin, Charlotte 40

eating *see* food
EBTA *see* Equine Behaviour and Training Association
elderly horses 99
    arthritis 103–4
    cognitive dysfunction syndrome (CDS) 105–6
    dentition 100–1
    digestion 101
    eyesight 104–5
    hearing 105
    pituitary pars intermedia dysfunction (PPID) 106
    separation anxiety 108
    shape 101–3
    sleep 106–7
    social status in herd 107
    euthanasia 110–11
        aftermath 113–14
        free bullet from a pistol 112–13
        lethal injection 112
    tips for owners 114–15
    welfare 108–10
Elizabeth, the Queen Mother 53
endurance riding 92
enrichment 21
    chart 24
    embedding 23–4
    food-based 20–1
        apple bobbing 21
        herbs 21
        licks 21
        root vegetables 20–1
    non-edible
        consistent carer 23
        field enrichment 23
        in-hand walks 23
        novel objects 22–3
        rolling 23
        routine tying 23
        rubbing mats 21
        stable toys 22, 27
        tea breaks 23
        toy box 21–2
epigenetics 33–4
equestrianism 79–80
    behaviour, Bedouin style 94–6
    behaviour in disciplines
        dressage 84–6
        driving 92–3
        hacking/endurance 89–92
        jumping 86–9
        polo 93–4
    behaviourally-minded 80–1
        preparing for competition 83–4
        travelling 81–3
Equicentral System 28–9
equid ethogram 5–6, 80
Equine Behaviour and Training Association (EBTA) 21
euthanasia 110–11
    aftermath 113–14
    methods
        free bullet from a pistol 112–13
        lethal injection 112

Fédération Equestre Internationale (FEI) 116
FEI *see* Fédération Equestre Internationale
Ferne Animal Sanctuary (Somerset) 72–3
Five Domains 10–11
Five Freedoms 9–10
Five Needs 10, 188
foals
   birth/environment 42–3
   bond with mare 43–4
   orphans 46–7
   weaning 47–9
food 4
   eating disorders 29
   elderly horses 100–1
   food-based enrichment 20–1
     apple bobbing 21
     herbs 21
     licks 21
     root vegetables 20–1
   natural eating behaviour 18–21
     browsing 17, 18
     hard feeds 20
     providing forage 18–20
   nutrition and behaviour 24–6
   patch foraging 18
   rescue/rehabilitation diet 141
   social needs 14

genetics 33–4, 54

habituation 171
hacking 89–92
   defecation 91
   separation anxiety 91
   startle response 90–1
HBCA *see* Human Behaviour Change for Animals
health
   arthritis 103–4
   cognitive dysfunction syndrome (CDS) 105–6
   dentition 100–1
   digestion 101
   diseases that cause/contribute to behaviour problems 163–5
   effects of flying 83
   eyesight
     ageing horses 104–5
     far-sightedness (hyperopic) 17
     short-sightedness (myopia) 17
   gastric ulcers 48
   hearing loss 105
   link between pain/behaviour problems 159–61
   pain issues 55
     acute 160
     chronic 160, 161
     common examples 160
     evasive behaviours 161
     facial expressions 160–1
     hiding/covering-up 161
     self-defence behaviours 160–1
     side-effects 161
   pituitary pars intermedia dysfunction (PPID) 106, 163–4
   separation anxiety 91, 108
   stress/behavioural problems link 165–6
   striving for balance 166–7
     coping strategies 167–9
     goal frustration 167
     inelastic demands 167
     maintenance behaviours 167
   working animals 189–90
   see also diseases
Healthy Land, Healthy Horses 28
herd management 150–1
   introduction 151–3
Hill Pony Improvement societies 40
human behaviour change 197
   consider wider change 207
     alternative technologies 210
     community engagement 209–10
     incentive-based schemes 209
     legislation/enforcement 207–8
     raising awareness 210
     regulation/control 208–9
   considering change 198
   culture 203
   education 203
   four principles
     change is a process 198–200
     change must be 'owned' 203–4
     environment influences change 202–3
     understanding psychology 200–2
   problem-solving 204–5
     finding the cause 205
       learning 206
       normal behaviour 206
       owner-horse relationship 207
       physiology 206–7
   psychology
     confirmation bias 200–1
     confrontational vs empathetic communication 200, 201
     elephant and rider model 201–2
     righting reflex 200, 201
     transactional analysis 200, 201
Human Behaviour Change for Animals (HBCA) 200, 203
hunting 88–9

IAABC *see* International Association of Animal Behaviour Counsellors
International Association of Animal Behaviour Counsellors (IAABC) 1
International Society for Equitation Science 85

jumping
    context specificity 87
    cross-country 86, 87–8
    habituation 86–7, 89
    hunting 86, 88–9
    point-to-point racing 86
    riding instruction 134
    shaping 88
    show jumping 86–7
    steeple-chasing 86
    working hunter showing classes 86

learning 62
    associative 62, 63
    classical conditioning 63
    counter-conditioning 63
    extinction/flooding 64–5
    habituation 63
    learnt helplessness 64
    non-associative 62–3
    operant conditioning 63, 76
    punishment 63–4
    reinforcement 63
        case study 65–6
        continuous 64
        non-contingent stimulus 64
        variable schedule of 64
    riding instruction theory 126–9
    sensitisation 63
    see also training
least invasive, minimally aversive (LIMA) approach 1
legislation 207–8
LIMA *see* least invasive, minimally aversive
Lockwood rehoming centre (Surrey) 155
long reining/long lining 93

mares
    bond with foal 43–4
    breeding
        free-range 37
        hand breeding 40
        man-made interventions 39
        pasture/paddock 40
        relationship with stallions 37–9
    foaling process/environment 42–3
    pregnancy 41–2
    in sport 40–1

movement
    after confinement 179–80
        station-focussed walking 180
        target stick walking 180–1
    benefits 16
    ethics of riding 17–18
    free-ranging 16
    necessity of 16
    socialisation 17
    welfare concerns 16–17

natural horsemanship 53, 67–9
needs
    basic 6
    elderly horses 109
    identifying 4–5
    priority of 6
    social
        eating/drinking 14
        human bond 81
        interactions 14
        rescue/rehabilitation 141
        stabling 13–15
        time element 6
negative reinforcement 67–9, 139

overshadowing 172–4

Pavlov, Ivan 171–2
Pillow Post 21
pituitary pars intermedia dysfunction (PPID) 106
polo 93–4
positive reinforcement 70–1, 147–8
PPID *see* pituitary pars intermedia dysfunction

quality of life 10

regulation 208–9
rescue and rehabilitation 136, 157–8
    behaviour in mind 150
        managing a changing herd 150–3
        matching people/equines 155–7
        rehoming 153–7
    environmental element 140–1
        diet 141
        movement 141
        social 141
        stimulation 141–2
    first few days 137–8
    giving good experience 138–40
    human element 142
        realistic expectations 144–6
        scientific approach to training 143–4
    negative reinforcement 139

restraint policy 139–40
safety issues 142
training element 147
positive reinforcement/rewards 147–8
successive approximation/shaping 148–50
riding, ethics of 17–18
riding instruction 116–17, 134–5
attention 125–6
attitude 119–20, 132, 133
avoid conflict between rider/instructor 132
benefits 117–18
client satisfaction 118
control 129–30
enthusiasm over progress 132
equine body language 122–5
learning theory 126–9
notice motivations 132, 133
responsibilities of instructor 134–5
reward smallest of improvements 132, 133–4
rider foundations 121–2
safe places 120–1
safety 118
set good example 132
setting goals 131–2, 133
teaching with behaviour in mind 130–1
avoid unwanted behaviour 125, 132, 133
examples 131–4
look for causes of unwanted behaviour 132
use slow shaping 132
welfare 118
what to teach 118–19

sedation drugs 176
self-defence 160–1
fiddle about 160
fight 160, 170
flight 160, 170
freeze 160, 170
shaping 55–9
avoidance of stress/fear 60–2
benefits 77
case study 59–60
critics of 58–9
linear plan 56
multi-faceted plan 57
plans for general training 150
rescue/rehabilitation 148–50
rules 57–8
Simpson, Heather 6
sleep 8, 10, 82–3, 106–7, 178, 179
Smith, Sharon 123
stables
barn system 13
bedding 20

design/layout 13, 14
wall systems 14
windows 14
group housing 14
inappropriate 17
meeting social needs 13–15
movement in/around 16–18, 80
outside access 13
using stable mirrors 15–16
visitors 14–15
stallions
abnormal behaviour 35–6
bachelor 36
free-living/free-ranging 36, 37
mating problems 39
pasture breeding/turnout with other stallions 36–7
points to consider 36
sexual advances 39
in sport 40–1
tending/courtship 38–9
stereotypies see abnormal behaviour
stress 26, 48, 141
acute 165
chronic 165–6, 167, 168, 170
management during confinement 177–9
and vets 170–6
successive approximation see shaping
Swiss National Stud 20, 37

Telatin, Angelo 85–6
Tersztyanszky, Colonel L. 85
Thomas, Colin 35, 40
Three F's 42
training 1, 13, 53
behavioural problems 54–5
clicker 75
clicker training 69, 71
controversies 77
early 45–6
experience of methods 66
harnesses/carts 192
reinforcing behaviour 65–6
negative reinforcement/positive punishment 67–9
positive 69, 71–2
positive vs negative 72–5
recommendations/ethics of positive vs negative 75–7
rescue/rehabilitation 143–4
science of learning 62–5
shaping 55–9, 148–50
avoidance of stress/fear 60–2
hoof-trimming example 59–60

training (*cont.*)
    traditional/natural horsemanship 67–9
    see also learning
Transtheoretical model of change 199, 209
travelling 81–3

UK Coaching Certificate (UKCC) 117
UKCC *see* UK Coaching Certificate
UNDP Cambodia Human Development Report (2016) 185

vets
    handling the horse 169–70
        chemical restraint 174–6
        counter-conditioning 171–2
        habituation 171
        overshadowing 172–4
    multi-modal approach 181–2
    treatment of medical problems/behaviour
        administration of medications 176–7
        box rest/reintroducing movement 177–9
        introducing movement after confinement 179–81

Waal, Frans de 127
Wadi Rum (Jordan) 94
Wake, Maisie 74, 75–6, 144
welfare 4, 6
    elderly horses 108–10
    Five Freedoms 9–11
    movement concerns 16–17
    riding instruction 118
    working animals 187–9
Welsh Pony and Cob Society 40
White, Jo 200, 203
working animals 184
    handling 190–1
        changes in behaviour 191–2
        equine vision 192
        in/out of harness 191
        training to accept harnesses/carts 192
    improving welfare 189–90
    numbers/distribution 186
    roles 186–7
    welfare issues 187–9

Xenophon 116